Environmental Science and Engineering

The ultimate goal of this series is to contribute to the protection of our environment, which calls for both profound research and the ongoing development of solutions and measurements by experts in the field. Accordingly, the series promotes not only a deeper understanding of environmental processes and the evaluation of management strategies, but also design and technology aimed at improving environmental quality. Books focusing on the former are published in the subseries Environmental Science, those focusing on the latter in the subseries Environmental Engineering.

Haoqing Xu
Editor

Proceedings of the 5th International Symposium on Water Resource and Environmental Management

Sustainable Development of Water Resource and Environmental Management

Springer

Editor
Haoqing Xu
School of Architecture and Civil Engineering
Jiangsu University of Science and Technology
Zhenjiang, Jiangsu, China

ISSN 1863-5520 ISSN 1863-5539 (electronic)
Environmental Science and Engineering
ISBN 978-3-031-31288-5 ISBN 978-3-031-31289-2 (eBook)
https://doi.org/10.1007/978-3-031-31289-2

This Springer imprint is published by the registered company Springer Nature Switzerland AG
The registered company address is: Gewerbestrasse 11, 6330 Cham, Switzerland

Preface

This book is designed to be the introductory work in the new Sustainable Development of Water Resource and Environmental Management Series. It provides an in-depth look at sustainable development and management in the water sector across.

In the face of current global changes, the availability and quality of water resources are under severe threat.

Indeed, in all sectors related to water resource management, sustainable development is important for present and future generations.

This book includes selected papers from the 5th International Symposium on Water Resource and Environmental Management (WREM 2022) and consists of themes pertaining to water resources and environmental management.

It provides readers with comprehensive information on the principles of sustainable water resource management, as well as recent advances, directions for future research, and policy development for sustainable water resource management.

The book will be useful for beginners, researchers, and professionals working in the area of water pollution and treatment, water engineering, and engineering structure.

Haoqing Xu, Ph.D.
Master Supervisor, Associate Professor
Deputy Director of Department of Geotechnical Engineering
School of Architecture and Civil Engineering
Jiangsu University of Science and Technology
Zhenjiang, Jiangsu, China

Contents

1 **Characterization of Groundwater Contamination in the North Lebanese Aquifer** 1
Wahib Arairo, Milad Khatib, Marianne Saba, Hassane Kassem, and Joseph Absi

2 **Spatial–temporal Variation of Vegetation Cover in the Dadu River Basin and the Impact of Cascade Hydropower Development** 17
Xiaoyan Shi, Yubin He, and Xiang Huang

3 **Research on Installation Height Optimization Technology of Energy Dissipation Box of Trailing Suction Hopper Dredger** 33
Hui Sun, Jiacheng Luo, Yuchi Hao, and Jiaming Qu

4 **Application Research of Educational Management in Fluidized Soil Engineering** 47
Yuan Fang, Xinmiao Shi, Haoqing Xu, Pengming Jiang, Yiyan Lv, and Kaidong Zhang

5 **Assessment of Green Infrastructures Performance for Water Quality Management** 61
Qian Yu and Na Li

6 **Removal Rate of Pollutants in Water Body by MBBR Based on Folded Cylindrical Biological Carrier** 69
Jian Zhu, Hui Tan, and Rui Wang

7 **Salt-Water Intrusion Prediction Using an Artificial Neural Network in the Bouregreg Estuary (Morocco)** 85
S. Haddout, K. L. Priya, A. M. Hoguane, and I. Ljubenkov

8 **Progress and Foresight in Wetland Environmental Science Research: An Econometric Analysis Based on the Knowledge Map of the WOS Core Database Literature** 95
Feiyan Huang, Yong Zhang, and Yunyao Ma

9 **Calculation for Resistance Loss of Fine Dredged Slurry in Long-Distance Straight Pipeline** 111
Rui Wang and Guoliang Yu

10 **Simulation Analysis of Sediment Corrosion of Mixed Transport Pipeline Based on FLUENT** 127
Minglei Xia and Guanglei Lv

11 **The Comparative Assessments of Growth Status of Sugarcane by TOPSIS Methods and Tissue Elements Distributions on Farmland of Heavily Polluted Environment** 143
Kaixian Zhu, Guiping Xu, Xiaofei Wang, Hexiao Chen, and Shiyang Yan

12 **Analysis of Water Demand Forecasting System Based on GIS Water Diversion in Yellow River Irrigation Area** 155
Wei Wang and Zhe Liu

13 **Biodiversity Associated to Floating Wetland Islands** 163
C. S. C. Calheiros, M. Ilarri, and S. I. A. Pereira

14 **Dynamic Evolution of Landscape Pattern in Hanzhong Pingchuan Section of Hanjiang River in China** 175
Xiaomei Kou, Qi Li, Hong Zhao, Yongxiang Cao, and Le Niu

15 **Smart Monitoring of Constructed Wetlands to Improve Efficiency and Water Quality** 189
Henrique Pinho, Manuel Barros, André Teixeira, Luís Oliveira, Pedro Matos, Carlos Ferreira, and Dina Mateus

16 **Experimental Study on Dehydration of Dredged Sediment from Urban Inland Rivers and Lakes and Analysis of Economic Benefits** 199
He Bai, Jiachen Zeng, Bing Guo, Yuchi Hao, Xiaowei Yan, and Zheng Lu

17 **Simultaneous Determination of 16 Metallic Elements in Surface Water by ICP-MS with Quadrupole Collision Cell Technology** 213
Yin-Fu Wang and Liu Yang

18 Spatio-Temporal Variation Characteristics of NPP in Weihe Watershed and Its Response to Environmental Factors Based on the CASA and CA–Markov Model 225
Lixia Wang, Feiyan Pan, Mingshuang Zhang, Zhao Liu, Shuangcheng Zhang, and Jinling Kong

19 Determination of Water-Soluble Cl^- and SO_4^{2-} in Soil by Chemical Analysis and Ion Chromatography 243
Liu Yang and Yin-Fu Wang

20 Hydrological Change and Probable Ecosystem Impacts Under a Climate Change in the Kunhar River, Pakistan 253
Shan-E-Hyder Soomro, Xiaotao Shi, Jiali Guo, Yinghai Li, Caihong Hu, Zhu Chunyun, Jiahui Gu, Ao Li, Shenghong Liu, Yu Guo, and Nida Rabab

21 Evaluation of Innovation and Entrepreneurship Education Based on Fuzzy Comprehensive Evaluation: A Case Study of Water Management 263
Jie Dai, Haoran Wei, Hengbin Yin, Chengqing Su, Jian Chu, and Yan Tu

Chapter 1
Characterization of Groundwater Contamination in the North Lebanese Aquifer

Wahib Arairo, Milad Khatib, Marianne Saba, Hassane Kassem, and Joseph Absi

Abstract Water is abundant on the earth's surface and underground. There is only 1% available to meet the needs of the population. Many groundwater resources' water quality has been impacted by water quality destruction and damage caused by river pollution. The North Lebanese Miocene Aquifer affected due to Karst topography by Kadisha-Abou Ali in North Lebanon is evaluated in this study. The river flows west, eventually reaching the Mediterranean Sea, and its basins well known for its limestone formations. Human agglomeration and activities (primarily industrial) are concentrated on the riverbanks of the Kadisha-Abou Ali watershed. Thereby the study aimed at assessing the groundwater from samples taken from the source of the river towards the sea to analyze the impact of pollution with population growth on the lower banks of the river, etc.... Thus, measuring the chemical, and physical parameters of all the tested samples, this study will determine the groundwater quality of the Miocene aquifer in northern Lebanon. The results of the sampled locations revealed a polluted pattern that was affected either rural activities such as olive culture in the northern part or suburban-related activities in the region of Tripoli at the bottom of the aquifer. Thus, a variety of factors have affected the fractured limestone aquifer, some of the factors included seawater intrusion, uncontrolled use of pesticides, and the overgrowth of the population.

Keyword Groundwater quality · Groundwater pollution · Chemical parameters · Physical parameters · Seawater intrusion

W. Arairo · M. Khatib (✉) · M. Saba · H. Kassem
University of Balamand, Koura, Civil and Construction Technology Department, Koura, Lebanon
e-mail: milad.khatib@isae.edu.lb

M. Khatib
ISSEA-Cnam, Civil Engineering Department, Zkak Blat Maurice Barres, Beirut, Lebanon

J. Absi
Université de Limoges, IRCER, UMR CNRS 7315, 12 Rue Atlantis, 87068 Limoges, France

H. Xu (ed.), *Proceedings of the 5th International Symposium on Water Resource and Environmental Management*, Environmental Science and Engineering,
https://doi.org/10.1007/978-3-031-31289-2_1

1.1 Introduction

Water is a vital component of life's sustainability; it contributes significantly to the expansion and development of towns and cities' environmental health. Water is used for more than drinking, cooking, and personal hygiene. Water is typically required in large quantities for industrial, commercial, and agricultural purposes. The pollution take place anytime dangerous materials, typically toxins or microbes, pollute the water, reducing its quality and rendering the water poisonous to users and/or the ecology.

Lebanon is located in the Eastern Mediterranean region, bordered by the Mediterranean Sea on the west. Although most of Lebanon receives a lot of rain when compared to its dry surroundings, certain areas in northeastern Lebanon get very little because of the rain shadow cast by the western mountain range's high peaks. The country has 17 year-round rivers, as well as hundreds of springs and 23 seasonal rivers. Data from a Miocene aquifer flowing through the Kadisha-Abou Ali River in northern Lebanon were collected for this study (Fig. 1.1).

The river flows west, eventually reaching the Mediterranean Sea. Human settlement and, primarily, industrial activity are concentrated along the riverbanks in the Kadisha-Abou-Ali Basin. The area has a Mediterranean character due to variations in topography (dramatic changes in topography just a few kilometers from coast to

Fig. 1.1 Map of the Kadisha-Abou Ali river

mountains) and climate (rain and snow in winter, drought in summer). The Kadisha-Abou Ali is a perennial river with high currents due to the steep slope, with maximum flows in the spring and minimum flows in the autumn.

Groundwater contamination is determined by several factors, including groundwater travel time in the aquifer (recharge over stored volume), contaminant quantity, contaminant transport, and contaminant attenuation zone. Natural groundwater recharge can occur because of direct infiltration (water intrusion from the surface into the subsoil) or subsequent infiltration. The underlying aquifer or saturated zone discharges runoff or seepage into the basal unsaturated zone (Scanlon et al. 2006). This formation is influenced by hydro-meteorological factors (precipitation, intensity, duration, and so on), surface hydrogeological conditions (topography, geology, soil science, and so on), and land use or cover (Siebert et al. 2010). In general, the influence of changing climate on surface water vs aquifer has been given significant attention, due to weak knowledge of worldwide impact of this change on aquifer (Green et al. 2011; Jackson et al. 2011; Holman et al. 2009; Mileham et al. 2008; IPCC 2007).

This is due, in part, to a data shortage, the complex and highly variable spatiotemporal response of aquifer reaction to changes in climate, as well as the difficulties of estimating underground aquifers (Green et al. 2011; IPCC 2007). Since correctly quantifying such influence is challenging, research have mostly focused on refill levels. Because it is difficult to quantify, accurately, the impact of climate change on groundwater storage, studies have focused primarily on recharge rates in watersheds, implying that recharge rates may be declining. There are some that point to potential recharge rate increases (Herrera-Pantoja and Hiscock 2008; Kruger et al. 2001), while others point to potential recharge rate decreases (Jyrkama and Sykes 2007; Kovalevskii 2007). There are many unknowns about climate change's effects on groundwater, but crop growth has an impact on both of direct and indirect aquifer refill increases. Both direct and diffuse potential groundwater recharge increases. Aquifer is inextricably linked hydraulically to the land beneath it. It is vulnerable to anthropogenic factors that have an impact on the land surface.

The ground in the study area is made up of highly fractured limestone soil. Subsoils can influence contaminant movement. Fine clay with low permeability can act as a groundwater protection layer. It is regarded as a barrier that prevents contaminants from moving. Sand and gravel, on the other hand, have a high permeability. It allows pollutants easy access to the water table and provides opportunities for pollutant dispersion among the pore spaces. Contaminant transport is greatly influenced by topographic elevation. Saba et al. (2016) evaluated the groundwater quality in Northern Lebanon due to saline water intrusion in previous work. Both of chemical and physical parameters were collected over the course of two years during different seasons. Because of the visible abilities in displaying the degree of pollution in the researched region, the outcomes were reviewed through Geographic Information System. (GIS). The Kadisha-Abou Ali River Basin in northern Lebanon is a symbol of water stress and increased consumption of water resources. Other forms of resource access, such as water table drilling, are mobilized because of resource consumption. Indeed, some villages in the region are experiencing severe water shortages, forcing rural residents to rely more and more on groundwater, and excavation is indiscriminate. The Kadisha-Abou Ali Basin is an important research target due to its increasing water use in water scarcity situations. Intensive groundwater development by the public and private sectors, as well as the public, depletes karst aquifer water resources in catchment areas.

From permitting to mining operations completion, the decision-making process for underground well good drilling remains a complex issue, sometimes lacking a scientific basis and sometimes beyond government control (e.g. indiscriminate wells).

Geologically, strata dating from the Cretaceous to the Quaternary distinguish the researched region. The coastal areas of this area are dominated by sands and sand loams that are partly alluvial in origin. The southern region dates from the Miocene epoch. The Miocene Limestone Aquifer is mostly composed of clastic and marl limestone "m2a," with thicknesses ranging from 250 to 300 m. A 150 m "mcg" thick conglomerate partially covers this aquifer. These calcite-consolidated conglomerates behave similarly to karst aquifers with underground channels in the underlying Miocene aquifer.

Groundwater in this basin flows primarily from the northwest to the Mediterranean Sea. Sir EL Danniye, a Cretaceous parallelogram aquifer separated by a Jurassic aquifer, occupies the eastern part of the Abou Ali basin. It is estimated to be 800 m thick. Groundwater in this aquifer flows generally southwest to northwest. 56% of the Abou Ali watershed has very high subsurface infiltration due to large fractures, which facilitates water infiltration. As the water and sanitation sector in Lebanon is facing many challenges due to the economic crisis, the growing need for clean water high increased. As the public water sector is almost absent, the increase in clean water overloaded the consumption and weight of use of groundwater. As such, the population is either relying on the use of licensed or unlicensed wells to attain the need for potable water. The main objective of this study is to collect aquifer samples selected comprehensively and selectively based on data facilities, and wells location to adequately integrate the water quality approach and assess the effects of seawater intrusion on the aquifer and gradually realize the human impact on aquifer quality.

1.2 Methodology

The project was guided by two study objectives:

i. To what degree are human activities damaging aquifer, which is their principal supply of drinkable water?
ii. How much is the geological formation and seawater intrusion affecting the groundwater of the North Miocene aquifer?

The water quality of the North Miocene Aquifer is assessed by analyzing aquifer sampling from the upper area near Bsharri to the coastline (Tripoli City). The main source of concern for water quality is catchment development, particularly near the shoreline, as well as overpopulation in Lebanon's second-largest city, which compromises groundwater quality. As a result, the focus of this study will be on assessing several groundwater locations (private wells) along the Abou-Ali River's stream from the upper to the lower part of the river to test the groundwater quality. The water samples were collected manually from nine different locations (Fig. 1.2 and Table 1.1) in North Lebanon, covering the upper, middle, and lower parts of the Miocene Watershed. Samples were taken using two different bottles, one polyethylene to cover the chemical testing and a sterilized cup to test the microbial parameters. The physical characterization was done onsite such as temperature to avoid any errors since the samples will be carried in a cooler and thus might affect the properties. Table 1.1 shows the sample locations and ID numbers.

The samples procedure included characterization of the physical, and chemical testing that was performed. The results were compared to the LIBNOR standard (LIBNOR 2010) and the standards of the Environmental Protection Agency (EPA) (U. S. Environmental Protection Agency (EPA). 1994). The parameters measured the concentrations of the main water quality parameters. They depict the water samples' conditions at the collection time, and can increase or decrease in proportion to the amount of runoff. The following parameters were evaluated as indicators of drinking water quality. The sample was characterized in three major steps. To begin, physical characterization was performed on-site using the Horriba machine (Fig. 1.2), which measured Total Dissolved Solids (TDS), Dissolved Oxygen (DO), Turbidity, pH, temperature, and Electrical Conductivity (EC).

This analysis was performed to avoid errors caused by fluctuations in the results and to obtain a direct idea about the physical properties of the sample water. After that, a sample of the 2 L polyethylene water bottle was taken to the laboratory to be chemically characterized. The parameters covered included the major anions and cations required for the chemical analysis of the water. Calcium, magnesium, sodium, potassium, nitrate, nitrite, and chloride were all measured in the ions analysis. The Ion chromatography instrument (Fig. 1.3 at the University of Balamand laboratory) is used for water chemistry analysis and measures the concentrations of major anions and cations in the ppm (mg/L) range.

1.3 Results and Discussion

The hydro-chemical, and hydro-physical analyses of groundwater samples taken from the aquifer and its surroundings revealed a variety of water patterns. The various parameter concentrations measured (Na^+, K^+, Mg^{2+}, Ca^{2+}, $SO_4{}^{2-}$, Cl^-, $NO_3{}^-$ and $NO_2{}^-$) during the sampling campaign revealed a general increasing pattern from west to east, paralleling the direction of groundwater flow, which is eastward.

The high concentrations observed are the result of over-pumping, which is causing either the arrival of deeper saline waters or the geological sedimentary formations of northern Lebanon. To begin, the pH values were depicted to determine whether they

Fig. 1.2 Satellite map of the Kadisha-Abou Ali river

Table 1.1 Sample locations

Designation	Name of location	Latitude	Longitude
H1	Ouadi Qadisha	342,839	359,221
H2	Hadath el Jebbeh	342,839	359,221
H3	Bsarma	343,881	358,674
H4	Dam and Farez (Tripoli)	344,204	358,438
H5	Abou Samra (Tripoli)	344,318	358,306
H6	Mallouleh Street (Tripoli)	344,612	358,504
H7	Bissar Street (Tripoli)	343,892	358,416
H8	200 Street (Tripoli)	344,359	358,293
H9	Maarad (Tripoli)	344,351	358,324

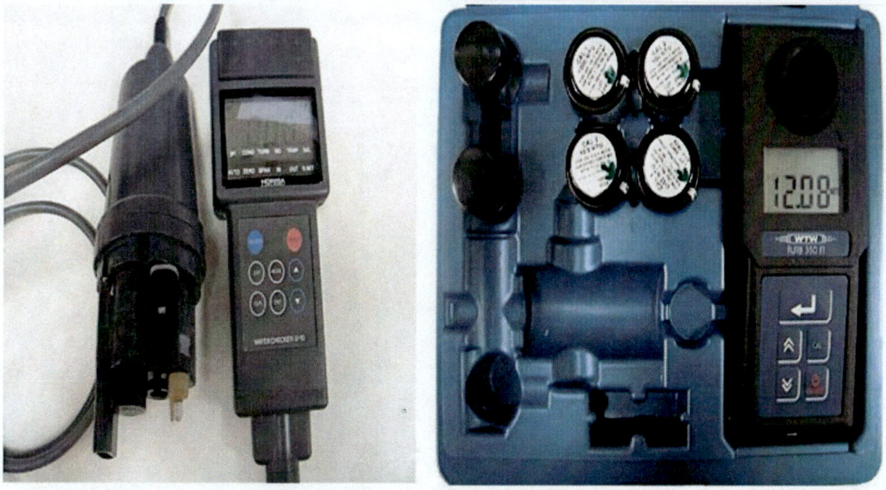

Fig. 1.3 Horriba apparatus used for the physical characteristics of the water

were within the Lebanese Standards (LS) and World Health Organization (WHO) standards (between 6.5 and 8). The results of the study revealed pH levels that are all within the normal range (Fig. 1.4).

Temperature variations were observed within the sample locations, with all samples having values greater than the set standard of 18 °C. This fact suggests that contamination, as well as other factors such as urbanization and the development of subsurface infrastructure, may be to blame for the rise in temperature (such as sewage systems). The warmth of shallow water aquifers will almost certainly rise rapidly because of this subsurface thermal pollution.

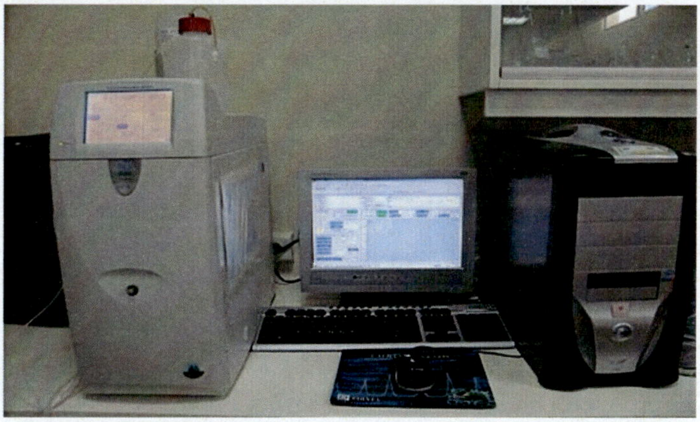

Fig. 1.4 Ion chromatography

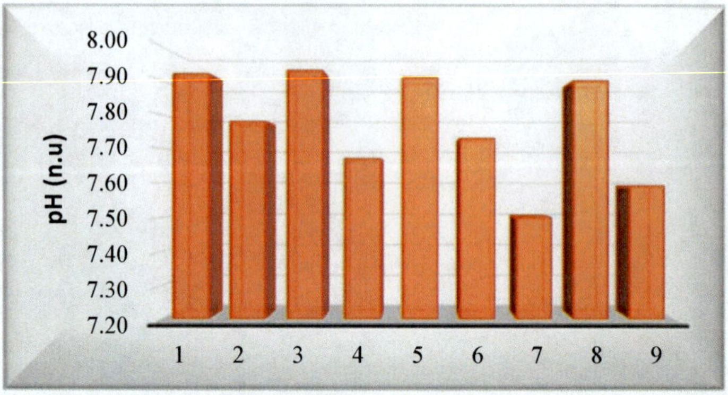

Fig. 1.5 pH values for the nine locations

The samples revealed a general turbidity trend; to achieve higher values, more samples were collected near the coastline. Furthermore, the values exceeded the standard implied by both the LS and WHO standards, which set a maximum level of 5 NTU (Fig. 1.6). This high level could be explained by water withdrawal from a very shallow depth, which increases the presence of dissolved solids and thus the turbidity of the water (Water is extracted at a 150 m to 250 m depth).

The dissolved oxygen levels in groundwater should be less than 8 ppm to be considered safe (referring to LS and WHO recommendations). All of the sample values indicated values above the normal limit (Fig. 1.7). These high levels are due to the usage of pesticides and fertilizer for agriculture activities in the upper region of the river.

Electrical conductivity (EC) and salinity values follow a general pattern: The closer the samples are to the sea, the higher the values, and the further the samples are

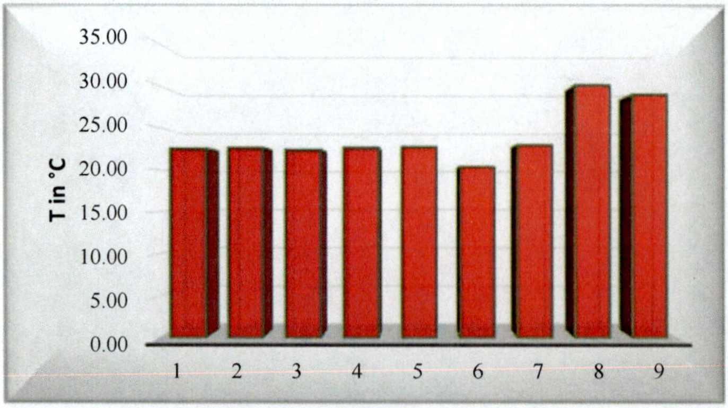

Fig. 1.6 Temperature variations (in °C) for the nine locations

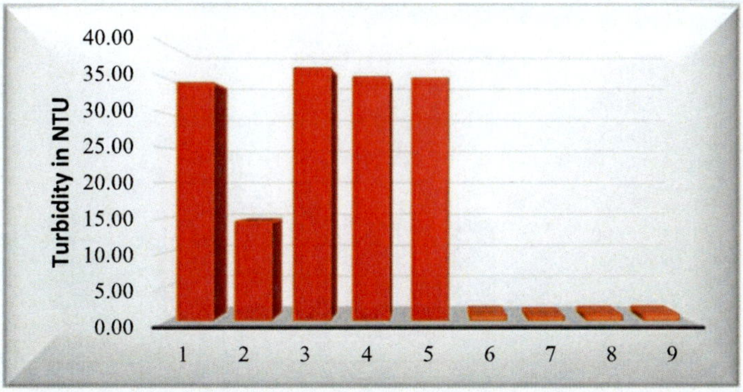

Fig. 1.7 Turbidity variations (in NTU) for the nine locations

from the sea, the lower the values. The conductivity (Fig. 1.8) and salinity (Fig. 1.9) values in Tripoli are the highest of any region up until they reach the sea.

This could be described by road runoff; coastal areas have significant quantities of inorganic chemicals due to seawater penetration, and the excessive immigration of Syrian refugees, all of which contribute to an increase in waste. However, all of the obtained values were within LIBNOR's specified limits (EC limits of 1500 μS/cm and Salinity limits of 0.05%).

The Total Dissolved Solids for three samples exceeded the LIBNOR limit (500 ppm), while another four samples were close to the limit (Fig. 1.10). This could be clarified by seawater intrusion, which raises the concentration of dissolved solids and minerals in groundwater.

The analysis of nitrate and nitrite concentrations in groundwater samples was performed to ensure the presence of contaminants in groundwater and to determine its feasibility. Except for H6, all samples have lower nitrate concentrations than the

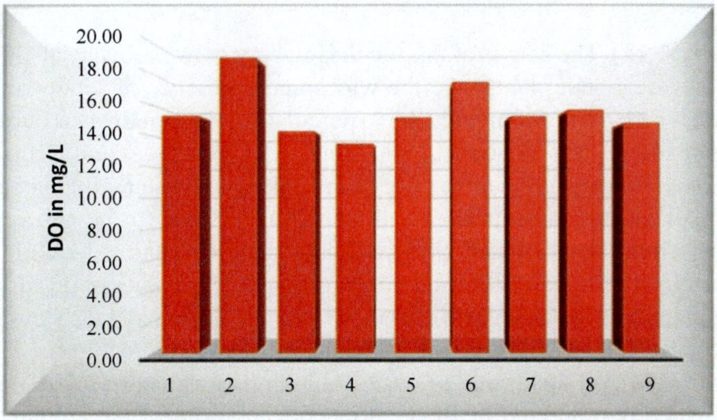

Fig. 1.8 Dissolved oxygen values

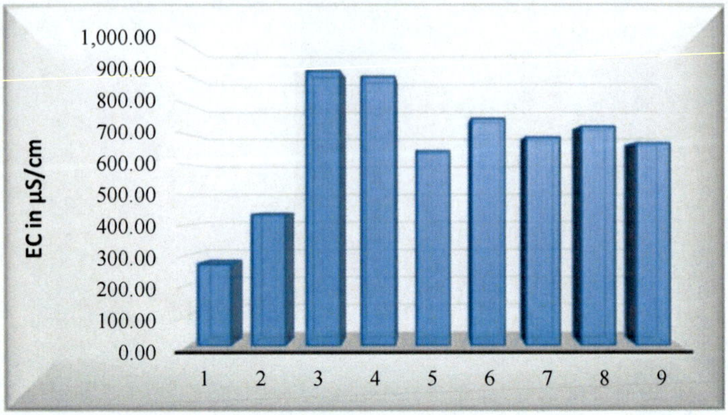

Fig. 1.9 Electrical conductivity values in µS/cm

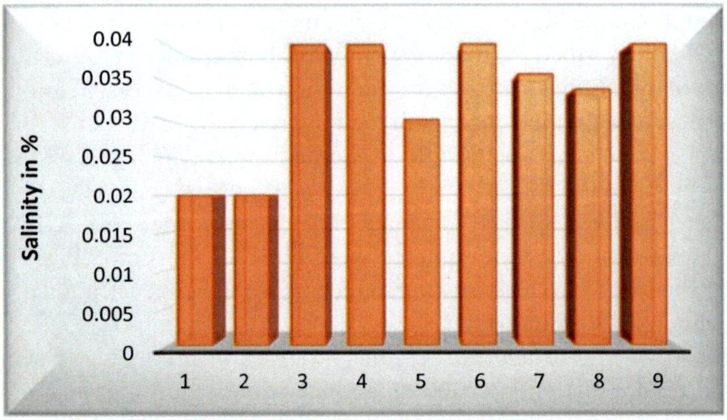

Fig. 1.10 Percentage of salinity variations

standard (Fig. 1.11). The causes of this are the improper sewage disposal, the scarcity of wastewater treatment plants, and the widespread presence of septic tanks in the Tripoli area, which allows contaminants to percolate into groundwater. Furthermore, before urbanization, Tripoli was an agricultural area. Pesticides and fertilizers may release trace amounts of elements into the soil and groundwater (Bijani et al. 2017).

In terms of nitrite concentrations, all values were found to be above the standard, which could be attributed to agricultural activities and a lack of a proper sewage system. In terms of chloride concentrations, the values showed a general increasing pattern as the samples were taken closer to Tripoli and the shoreline. The values exceeded the acceptance contaminant level (MCL) of 1 mg/L. (Fig. 1.12). This demonstrates yet again that seawater intrusion is a major problem along the coastline, which is supported by the geological formations of the ground. Because

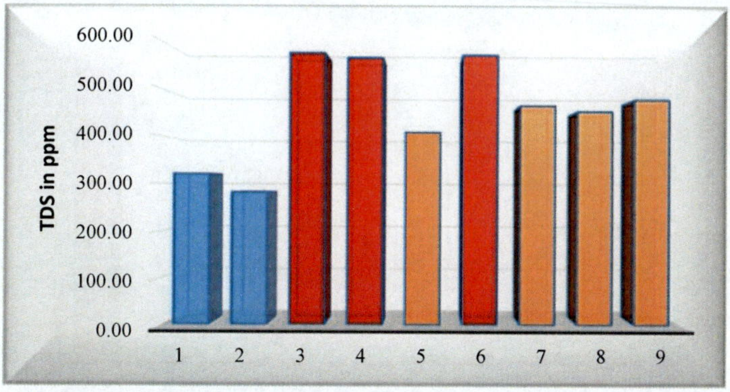

Fig. 1.11 Total dissolve solids values in ppm

of the porosity of the ground, chemicals such as chloride can percolate into the groundwater.

Sulfate has a maximum allowable level of (250 mg/L) and fluoride has a maximum allowable level of (1 mg/L). All sulfate and fluoride concentration levels were found to be within LS and WHO guidelines. When the chemical characterization is combined with the geological formation of the North Miocene Aquifer, the water chemistry can be studied in greater depth.

All the tested samples show that the cations values (Na^+, K^+, Mg^{2+}, Ca^{2+}) are within the limits.

A Piper diagram (Fig. 1.13) will allow a visual approach to transmitting the meaning of water chemistry to help with understanding the sources of the dissolved component salts in water (Masoud and Khamehchi 2019). The concentration of cations and anions in water, as well as the fact that the algebraic sum of cation and anion electric charges is zero, ensures the electro-neutrality of the dissolved salts.

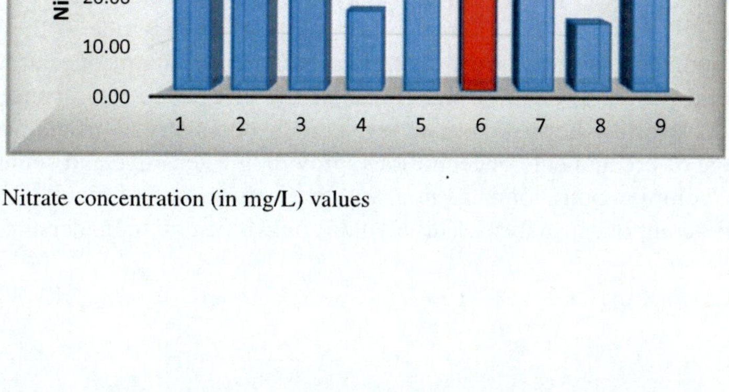

Fig. 1.12 Nitrate concentration (in mg/L) values

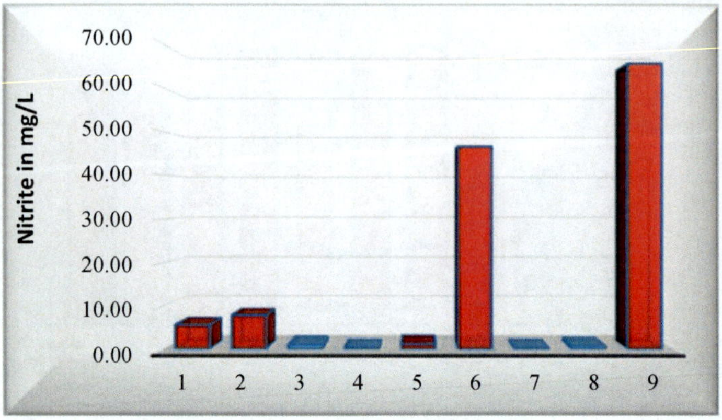

Fig. 1.13 Nitrite concentration (in mg/L) values

The Piper diagram of hydrogeological for the research region clearly demonstrates yet if the inorganic chemicals of the Abou Ali River and Miocene Aquifers are connected to the dominance of rainfall, rock, or evaporated crystallization type.

Samples having low TDS and significant $Na^+/(Na^+ + Ca^{2+})$ or $Cl^-/(Cl^- + NO_3^-)$ proportions (near to 1) are mostly found in the lower-right corner, showing that rainfall is the dominant factor. However, samples with mildly higher TDS and $Na^+/(Na^+ + Ca^{2+})$ or $Cl^-/(Cl^- + NO_3^-)$ proportions of 0.5 or less are generally found in the middle zone, showing rock dominance. Moreover, samples with extremely high TDS with high $Na + /(Na^+ + Ca^{2+})$ or $Cl^-/(Cl^- + NO_3^-)$ proportions are mostly found in the upper-right corners, showing evaporated crystallization, which reflects the effect of evaporation in dry environments.

Furthermore, the studied water surface with a $Ca^+/(Ca^+ + Mg^{2+})$ ratio larger than 0.5 and a high TDS content (upper right corner of the diagram) showing that it is of evaporative crystallization dominant origin. The aquifer samples are largely dispersed in the evaporated crystallization dominance region, with a slightly shift in the rock zone, illustrating that evaporated crystallization and wind erosion govern the chemistry of the water, implying that the groundwater is generally hyper-chlorinated and sulfated with magnesium because of the calcined formation of the aquifer.

The Stiff diagram is used to show how a water body's ionic composition changes over time and space. It facilitates the identification of ion-related waters from which a flow channel can be identified. Fig. 1.14 plots the equivalent concentrations of anions to the right and cations to the left of the central axis by connecting the points. When comparing Stiff diagrams of different water samples, the left side of the diagram indicates cation concentrations and the right side indicates anion concentrations.

As a point moves away from the center of the graph, the ionic concentration increases. As a result, the stiff diagram revealed the chemistry of groundwater over a wide range of component concentrations, providing a sensitive and sophisticated analysis of complex pollution sites in a way that is extremely difficult to replicate with a single-component analysis. This method could be useful for understanding and

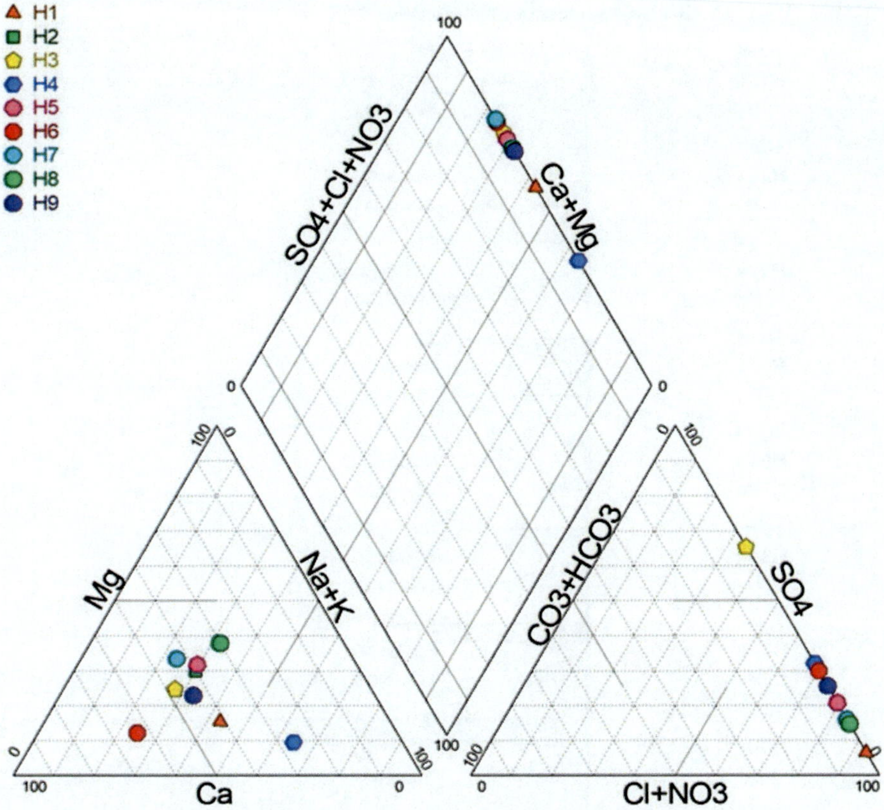

Fig. 1.14 Piper diagram of the conducted water samples

tracking mixing and chemical changes in uncontaminated environments. The results revealed geological formations and seawater intrusion, particularly in location H8, highlighting the importance of this tool in the water quality of the North Lebanese Miocene aquifer.

The Schoeller-Berkaloff diagram (Figs. 1.15 and 1.16) revealed extremely comparable patterns among groundwater samples, particularly the Piper pattern, confirming the same fingerprint between geological formations and groundwater quality. The sample H8, which is located more peripherally than the other samples, has a higher concentration of Na^+ and K^+ crosslinking, as seen in the Stiff diagram (Fig. 1.14).

1.4 Conclusion

All sedimentary rocks in the Tripoli region and its surrounding area are predominately sedimentary; with calcium carbonate, $CaCO_3$, as the main component of these rocks

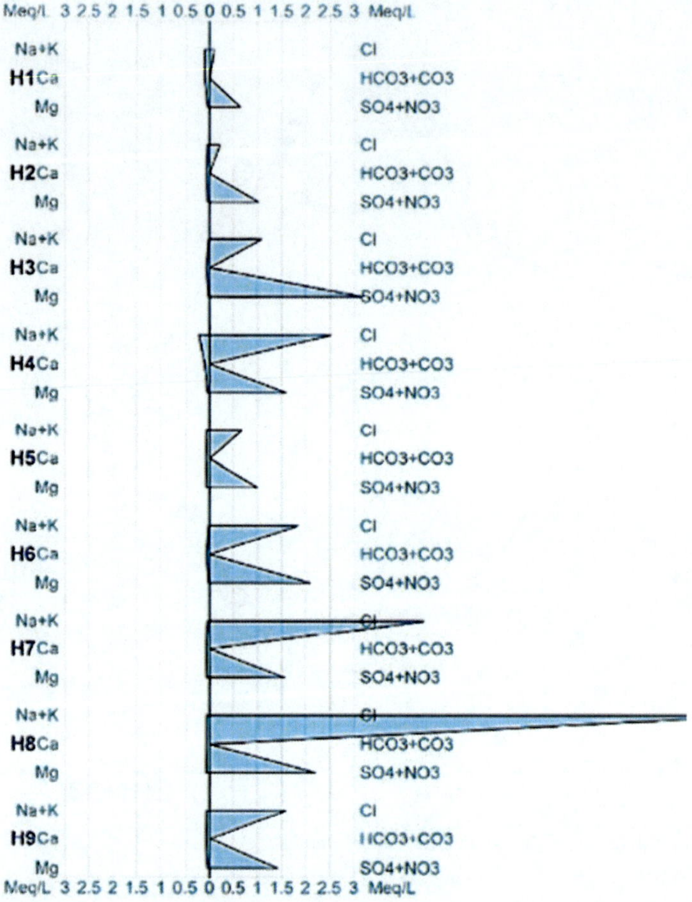

Fig. 1.15 Stiff diagram of the sampled groundwater

and additional minerals ($MgCO_3$; SiO_2) depending on the layer in which groundwater is available.

The chemical composition of the groundwater sampled varied due to variables that vary with elevation and geographical distance from the shoreline. It was discovered that the following factors influenced groundwater quality: Water temperature, type and solubility of different minerals in soil or aquifer, and distance from the sea (seawater intrusion).

The magnitude per each effect was governed in partly by the water's residency period in the diverse subterranean ecosystems. The upward movement of saltwater can cause high sodium and chloride levels from deeper bedrock in areas of high pumping. Thus emphasizing that the North Miocene groundwater contains detectable concentrations of chemicals related to three major aspects:

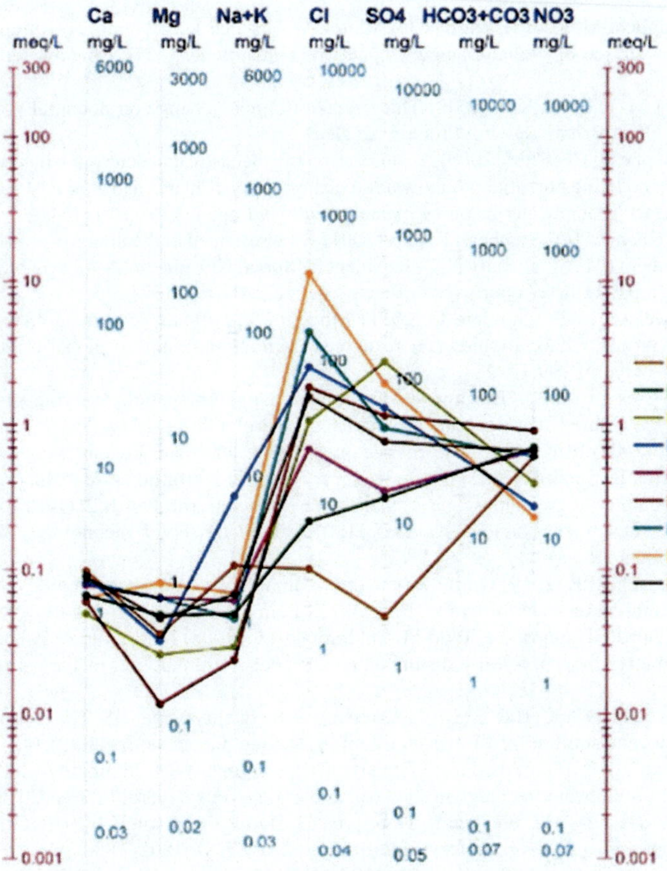

Fig. 1.16 Schoeller-Berkaloff diagram of the sampled groundwater

Geological settings: Both industrial and agricultural pollutants, in addition to decayed organic materials like animal sludge or sewage, are absorbed. Development of seawater intrusion on the Tripoli coastline.

A clear distinction between anthropogenic factors (like agricultural activities and over-pumping those results in seawater intrusion) and natural factors (like the geologic) was established by analyzing the hydrochemical and hydrophysical sets of results obtained from the seasonal variations.

References

Bijani M, Behbahani Bijani RM, Moghadasi J (2017) Predicting scale formation in wastewater disposal well of Rag-e-Safid desalting unit No.1. Desalination Water Treat 65: 117–124

Green TR, Taniguchi M, Kooi H, Gurdak JJ, Allen DM, Hiscock KM, Treidel H, Aureli A (2011) Beneath the surface of global change: impacts of climate change on groundwater. J Hydrol 405:532–560

Herrera-Pantoja M, Hiscock KM (2008) The effects of climate change on potential groundwater recharge in Great Britain. Hydrol Process 22(1):73–86

Holman IP, Tascone D, Hess TM (2009) A comparison of stochastic and deterministic downscaling methods for modeling potential groundwater recharge under climate change in East Anglia, UK: implications for groundwater resource management. Hydrogeol J 17:1629–1641

IPCC: Climate Change 2007 Synthesis Report (2007) An assessment of the intergovernmental panel on climate change. IPCC Plenary XXVII, Valencia, Spain, November 12–17 p 52 https://www.globalchange.gov/browse/reports/ipcc-climate-change-2007-synthesis-report

Jackson CR, Meister R, Prudhomme C (2011) Modeling the effects of climate change and its uncertainty on UK Chalk groundwater resources from an ensemble of global climate model projections. J Hydrol 399:12–28

Jyrkama MI, Sykes JF (2007) The impact of climate change on spatially varying groundwater recharge in the Grand River watershed (Ontario). J Hydrol 338(3–4):237–250

Kovalevskii VS (2007) Effect of climate change on groundwater. Water Resour 34(2):140–152

Kruger A, Ulbrich U, Speth P (2001) Groundwater recharge in Northrhine-Westfalia predicted by a statistical model for greenhouse gas scenarios. Phys Chem Earth Part B 26(1–2):853–861

LIBNOR (2010) Document Number: NL ISO TR 13530. Sector: Food Technology. TC: NL TC 147. ICS: 13.060.45

Masoud B, Khamehchi E (2019) Optimization and treatment of wastewater of crude oil desalting unit and prediction of scale formation. Environ Sci Pollut Res 26:25621–25640

Mileham L, Taylor R, Thompson J, Todd M, Tindimugaya C (2008) Impact of rainfall distribution on the parameterization of a soil-moisture balance model of groundwater recharge in equatorial Africa. J Hydrol 359:46–58

Saba M, Gerges N, Iaaly A, Carlier E (2016) Assessing water quality using GIS: The case of northern Lebanon miocene aquifer. Int J Environ, Chem, Ecol, Geol Geophys Eng 10(2):147–155

Scanlon BR, Keese KE, Flint AL, Flint LE, Gaye CB, Edmunds WM, Simmers I (2006) Global synthesis of groundwater recharge in semiarid and arid regions. Hydrol Process 20:3335–3370

Siebert S, Burke J, Faures JM, Frenken K, Hoogeveen J, Döll P, Portmann FT (2010) Groundwater use for irrigation—a global inventory. Hydrol Earth Syst Sci 14:1863–1880

U. S. Environmental Protection Agency (EPA) (1994) Water quality standards handbook. Chapter 6: Procedures for Review and Revision of Water Quality Standards. EPA-823-B-17-001. EPA Office of Water, Office of Science and Technology. Washington, DC, Accessed October 2022, https://www.epa.gov/sites/production/files/2014-10/documents/handbook-chapter6.pdf.

Chapter 2
Spatial–temporal Variation of Vegetation Cover in the Dadu River Basin and the Impact of Cascade Hydropower Development

Xiaoyan Shi, Yubin He, and Xiang Huang

Abstract The spatial–temporal variation characteristics of NDVI in the whole Dadu River basin and the affected area of the cascade hydropower during the 22 years were studied based on the SPOT/VEGETATION dataset and precipitation and temperature data from 1998 to 2019 using pixel by pixel one variable linear regression and correlation analysis. The results showed that (1) The mean annual NDVI of the Dadu River basin maintained fluctuating growth, with an increase rate of 0.00476/a ($p < 0.05$) from 1998 to 2019. (2) The NDVI growth trend of vegetation in all areas of basin during the study period was significant, with 92.11% of the pixels showing a growth trend and of which 82.18% was significant. (3) The correlation between NDVI and temperature and precipitation in basin showed obvious regional differences. In the downstream area near the Chengdu Plain, NDVI is negatively correlated with precipitation and positively correlated with temperature while NDVI is positively correlated with temperature and precipitation in the northern area of Jinchuan County. (4) Comparing the NDVI changes before and after the construction of the hydropower plant, it shows that the hydropower plant may lower the NDVI value in the early stage of construction, but the vegetation will recover rapidly with the completion of the plant and will not change the original change trend. The above results show that the NDVI changes are mainly influenced by meteorological conditions, and the negative impacts of the hydropower plant construction on the basin level are limited and recoverable.

Keywords Dadu river basin · NDVI · Spatial–temporal features · Cascade hydropower · Correlation analysis

---

X. Shi · Y. He
CHN Energy Dadu River Big Data Service Co Ltd, Chengdu 610041, China

X. Huang (✉)
CHN Energy Dadu River Basin Hydropower Development Co Ltd, Chengdu 610041, China
e-mail: huangxiang125125@163.com

H. Xu (ed.), *Proceedings of the 5th International Symposium on Water Resource and Environmental Management*, Environmental Science and Engineering,
https://doi.org/10.1007/978-3-031-31289-2_2

The development of hydropower plant s can have huge social and economic effects, but it also changes the structure and function of terrestrial and aquatic ecosystems to a certain extent. When hydropower plant s are built and impounded, they will inevitably inundate part of the land leading to the loss of upstream terrestrial habitats (Botelho et al. 2017), and the increase in upstream water surface will also create microclimates, which in turn will affect the structure and diversity of upstream vegetation (Zhang et al. 2020). In addition, hydropower plant construction and operation alter the natural flow of rivers and directly lead to changes in sediment hydrology, resulting in the loss of rapids and longitudinal watercourses, affecting historical channel dynamics and alluvial geomorphology downstream. In many cases, as these changes may have a noticeable impact on riparian vegetation, leading to changes in structure and distribution, decreases in species richness and invasion of exotic species (Liu et al. 2022a), therefore, there is a lot of interest in the overall ecological impact on the basin.

Normalized Difference Vegetation Index (NDVI) is constructed by using the reflection characteristics of vegetation in visible and near-infrared wavelengths, and is often used as the best biophysical indicator of vegetation growth status and cover change, which especially the time series NDVI can better reflect the temporal evolution and spatial variability of vegetation activities on land surface In particular, a time-series NDVI can better reflect the temporal evolution and spatial variability of vegetation activities on land surface (Chu et al. 2019; Peng et al. 2019), which has been widely used in vegetation change studies at different scales (Yang et al. 2019). For example, when Liu et al. studied the spatial and temporal variation of NDVI in various terrestrial ecosystems in China over the past 30 years, it was found that the vegetation activity in China has increased significantly over the past 30 years (Liu et al. 2018). Ma et al. studied the karst landscape area in southwest China and found that the response of NDVI to climate factors is spatially and temporally heterogeneous (Ma et al. 2021). Ouyang et al. found that the severely affected vegetation areas were concentrated within 5 km from the reservoir by comparing the interannual NDVI around reservoirs (Ouyang et al. 2010). Yi et al. showed that the construction of the dam trapped a large amount of water, sediment and nutrients in the upstream area, and the increase in air humidity and nutrients favored vegetation growth (Yi et al. 2019).

The environmental impact assessment of hydropower plant s in general tends to focus on a small area and a short time span, lacking long-term studies and judgments on the extent of disturbance and overall impact results at the whole basin level (Li et al. 2012), while the longer duration and higher degree of development of the Dadu River basin is a good case to explore this issue. In addition, the current remote sensing data for assessing terrestrial vegetation cover can basically cover the pre and post time periods of the Dadu River basin development, and the application conditions are available.

Based on the above reasons, this paper takes the Dadu River basin as an example and tracks the long time series data from 1998 to 2019 in terms of NDVI, precipitation, temperature and other latitudes to comprehensively analyze the vegetation cover changes and spatial distribution characteristics of the Dadu River basin and further investigate the impact of hydropower gradient development on the vegetation cover. The results of the study will reveal the impact of the Dadu River basin on vegetation cover. The results of the study will reveal the spatial–temporal variation characteristics of NDVI in the Dadu River basin and its influencing factors, and provide theoretical support for the development of ecological protection and sustainable development of the Dadu River basin.

2.1 Overview of the Study Area

The Dadu River is the largest tributary of the Minjiang River system in the upper reaches of the Yangtze River, originating from the southeastern foot of Goluo Mountain in Qinghai Province, with two sources: the eastern source is the Foot Muzu River and the western source is the Choshujia River, with the eastern source being the main source. The two sources from the mouth of the double river (i.e. Aba Malcom County, Baiwan Township) after the confluence began to be called Dadu River. Its latitude and longitude are between 99°42′E and 103°48′E, and its latitude is between 28°15′N and 33°33′N. The total length of the main river is 1062 km, with a natural drop of 4175 m, including 852 km in Sichuan Province, with a natural drop of 2788 m; the total river basin area is 77400 km^2 (excluding the Qingyi River), with an annual runoff of 47 billion m^3, including a basin area of 70,821 km^2 in Sichuan Province, accounting for 91.5% of the total basin (He et al. 2021).

Dadu River main stream hydropower planning is mainly power generation, taking into account downstream flood control, navigation, water supply, etc. The development of Dadu River main stream basin lasted for a long time from 1965 when Gongzui hydropower plant was built, through planning adjustments and optimization studies of downstream and local river development methods, the whole basin was developed with 28 stages, at present, among the planned hydropower plant s of Dadu River main stream, Monkey Rock, Changhe Dam, Jinjinping, Luding, Dagangshan, Longtoush, Waterfall, etc. Dagangshan, Longtou Shi, Waterfall Gorge, Shenxigou, Pillow Dam I, Shaping II, Gongzui, Tongjiezi, Shawan and Angu are built hydropower plant s, Shuangjiangkou, Jinchuan, Hard Liangbao, Pillow Dam II and Shaping I are under construction, and most of the remaining steps are in pre-feasibility or feasibility study stage. The construction and spatial distribution of the main-stream stages are shown in Fig. 2.1.

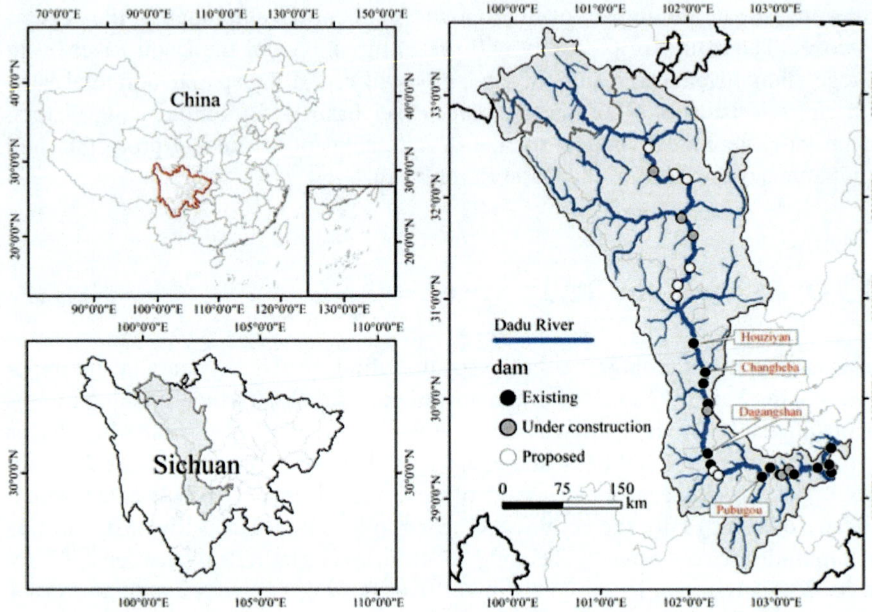

Fig. 2.1 Location of study area

2.2 Data and Study Method

2.2.1 Data Information

2.2.1.1 NDVI data

At present, NDVI time-series data obtained from satellite remote sensing images based on SPOT/VEGETATION and MODIS have been widely used in studies of vegetation dynamics monitoring, cover change detection, macroscopic vegetation cover classification and net primary productivity estimation in various scales (Liu et al. 2022b). The NDVI dataset used in this study is an annual vegetation index dataset generated based on the SPOT/VEGETATION (http://www.vito-eodata.be) decadal 1 km vegetation index data, using the maximum synthesis method to reduce the effects of the atmosphere, clouds, and solar zenith. index dataset (Holben 1986). This dataset effectively reflects the distribution and change of vegetation cover in spatial–temporal scales in all regions of the country, and is a very important reference for monitoring vegetation change status, rational use of vegetation resources and other ecological and environmental related fields of research. The data can be obtained from the Data Center for Resource and Environmental Sciences, Chinese Academy of Sciences (http://www.resdc.cn/).

2.2.1.2 Meteorological data

Regarding the relationship between vegetation growth and climate, temperature and precipitation are usually considered to be the most important climatic factors affecting vegetation dynamics. The annual spatial interpolation dataset of precipitation and temperature in China used in this study is based on daily observations of meteorological elements at more than 2400 plants nationwide, which is available from the Data Center for Resources and Environmental Sciences, Chinese Academy of Sciences (http://www.resdc.cn/). Spatially interpolated data of precipitation and temperature for each year from 1960 to 2021 generated by interpolation based on Anuspl interpolation software using smooth spline function on the basis of calculating annual values of each meteorological element (Interpolating mean precipitation using thin plate smoothing splines 1995). The annual average temperature unit is 0.1 °C, and the annual precipitation unit is 0.1 mm, and the spatial resolution is 1 km. The meteorological data were aligned with NDVI data by projection and row number alignment using ArcGIS 10.8 software.

2.2.2 Research Methods

2.2.2.1 Linear regression and significance test

Least squares was used to analyze the temporal rate of change of each pixel in the study area, thus reflecting the characteristics of the evolution of NDVI during the 22 years starting from 1998 to 2019.

$$Slope = \frac{n \times \sum_{i=1}^{n} i \times \mathrm{NDVI}_i - \left(\sum_{i=1}^{n} i\right) \times \left(\sum_{i=1}^{n} \mathrm{NDVI}_i\right)}{n \times \sum_{i=1}^{n} i^2 - \left(\sum_{i=1}^{n} i\right)^2} \tag{2.1}$$

where Slope is the slope of the NDVI regression trend, n represents the duration in years, and NDVI_i represents the corresponding NDVI value at year i. When Slope > 0, it indicates an increasing trend of NDVI in the study area during n years; on the contrary, it indicates a decreasing trend of NDVI; if Slope is close to 0, it means that the change is not significant.

2.2.2.2 Correlation analysis

Meteorological elements are important factors affecting vegetation cover, and the growth of plants and the composition of vegetation types cannot be separated from the role of temperature as well as precipitation. In order to obtain the pattern of NDVI changes with meteorological changes, two variables, precipitation and temperature, were selected as the unit of analysis, and the correlation between NDVI and these two meteorological factors was determined, so as to obtain the mode of action of climate on vegetation. The correlation coefficient is used to quantitatively determine the closeness of the correlation between the geographic factors, and the simple correlation coefficient is calculated as follows.

$$r_{x.y} = \frac{\sum_{i=1}^{n}(x_i-\overline{x})/(y_i-\overline{y})}{\sqrt{\sum_{i=1}^{n}(x_i-\overline{x})^2}\sqrt{\sum_{i=1}^{n}(y_i-\overline{y})^2}} \tag{2.2}$$

where, $\overline{x}$ and $\overline{y}$ are the average values of elements x and y, respectively. The value of simple correlation coefficient is $-1 \leq r_{xy} \leq 1$, $r_{xy} \geq 0$ means that the two judged are changing positively to each other, while if it is less than 0, it is showing a reverse change. The smaller the difference between the absolute value of the calculated data and 1, the greater the correlation, r_{xy} the closer it is to 0, indicating that the correlation between the elements is not strong. The significance test of NDVI and climate elements in different years was conducted using *t*-statistic, and the *t*-statistic was defined as.

$$t = \frac{r\sqrt{n-q-2}}{\sqrt{1-r^2}} \tag{2.3}$$

where r is the correlation coefficient, n is the number of samples, q is the order, and the statistic t obeys a t distribution with $n - q - 2$ degrees of freedom. By the t-value, the probability P-value corresponding to it is calculated.

2.3 Results and Analysis

2.3.1 *Overall Spatial Variation Characteristics of NDVI in Dadu River Basin*

2.3.1.1 NDVI variation trend

The results of using one-dimensional linear regression analysis and least squares method to fit the annual maximum NDVI mean values over 22 years are shown in Fig. 2.2, the annual NDVI mean changes in the Dadu River basin during the study period showed an overall fluctuating upward trend Slope of 0.00476/*a* ($p < 0.05$), 1998–2001, 2002–2006, 2008–2010, and 2012–2016, the NDVI of the Dadu River basin showed a rapid increasing trend, and the NDVI of the Dadu River basin in 2001–2002, 2006–2008, and 2010 showed a rapid increasing trend, 2006–2008, and 2010–2012 NDVI values showed a decreasing trend followed by a rapid recovery. The average value of NDVI in the Dadu River basin during the study period was 0.732, where the maximum value occurred in 2019 with 0.781 and the minimum value occurred in 1998 with 0.677.

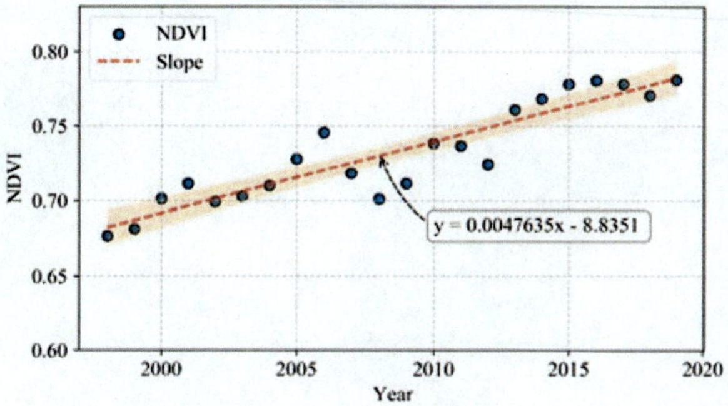

Fig. 2.2 Annual average of NDVI in the Dadu River Basin

The slope of each NDVI pixel by pixel was fitted to obtain the multi-year trend of NDVI spatial distribution, and the direction and rate of multi-year NDVI change in the basin were analyzed comprehensively. In, the NDVI change rates were divided into six categories: large decrease (<−0.008), medium decrease (−0.008 ~ −0.004), small decrease (−0.004 ~ 0), small increase (0 ~ 0.004), medium increase (0.004 ~ 0.008), and large increase (>0.008). The percentage of each change rate and spatial distribution as shown in Table 2.1 and Fig. 2.3. The percentage of the decreasing pixel only accounts for 7.89% of the total, and the large and moderate decreasing areas are mainly located in the eastern part of Xiaojin County and Markang City, the high altitude areas on both sides from Monkey Rock hydropower plant to Waterfall Gorge hydropower plant, the inundation area of Waterfall Gorge hydropower plant and around the downstream section from Gongzui hydropower plant to Hekou River, and the overall increasing trend of the basin NDVI, with the largest percentage of the moderate increasing area being 62.42%, followed by the small increasing area being 62.42%. The largest increase was 62.42%, followed by a small increase of 19.71%, and a large increase of 9.89%.

Table 2.1 NDVI change rate in the Dadu River basin from 1998 to 2019

Slope(/a)	Change rate	Number of pixel	pixel percentage (%)
<−0.008	Significant decrease	1084	1.42
−0.008 ~ −0.004	Medium decrease	1577	2.07
−0.004 ~ 0	Small decrease	3430	4.50
0 ~ 0.004	Small increase	15,038	19.71
0.004 ~ 0.008	Medium increase	47,629	62.42
>0.008	Significant increase	7548	9.89

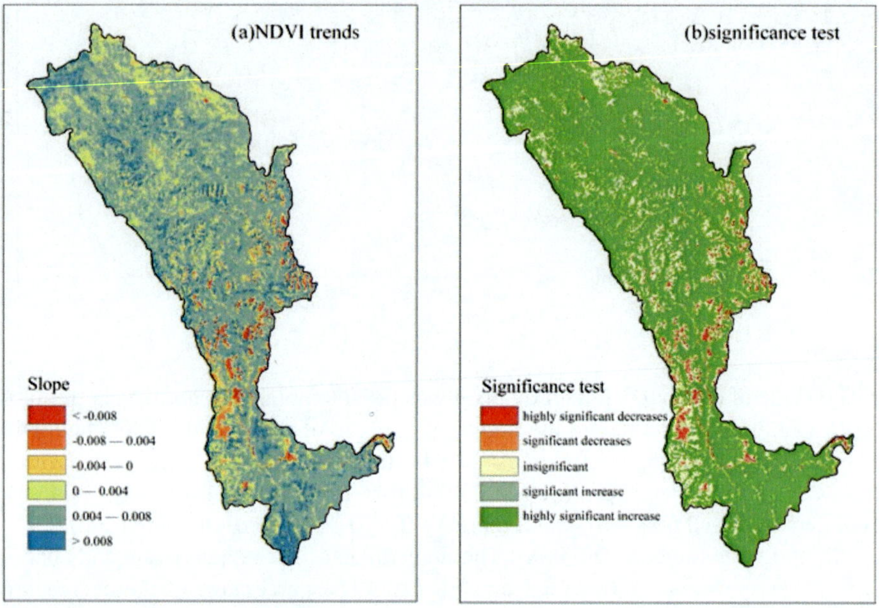

Fig. 2.3 Spatial distribution of NDVI trends and significance test

The significance test of pixel was conducted at $p = 0.05$ and $p = 0.01$ levels, and the results were highly significant for less than 0.01, significant for 0.01 ~ 0.05, and insignificant for more than 0.05. The results of are shown in Fig. 2.3. The majority of the areas in the basin showed a highly significant increase in NDVI, accounting for 76.01%, and 14.69% of the areas with insignificant changes, while the areas with significant and highly significant decreases accounted for only 3.14% of the basin area, with a similar distribution range as the areas with Slope < 0.

In summary, the NDVI of the Dadu River basin from 1998 to 2019 showed a significant increasing trend, and the average value of Slope was 0.00476/*a*. The NDVI of the basin only decreased in some high altitude areas from Luding to Jinchuan, but the rest of the areas showed a significant increasing trend.

2.3.1.2 Factors influencing NDVI changes

Research results show that regional climate change is an important cause of NDVI increase, while temperature and precipitation are important indicators of climate change (Hao et al. 2023). The annual mean temperature with precipitation and NDVI variation of basin in the study period is obtained by calculating the pixel average of the study area from year to year as shown in Fig. 2.4. The multi-year average temperature of Dadu River basin is 4.72 °C, the maximum annual average temperature occurs in 2006 at 5.26 °C, and the minimum value of 4.14 °C occurs in 2004.

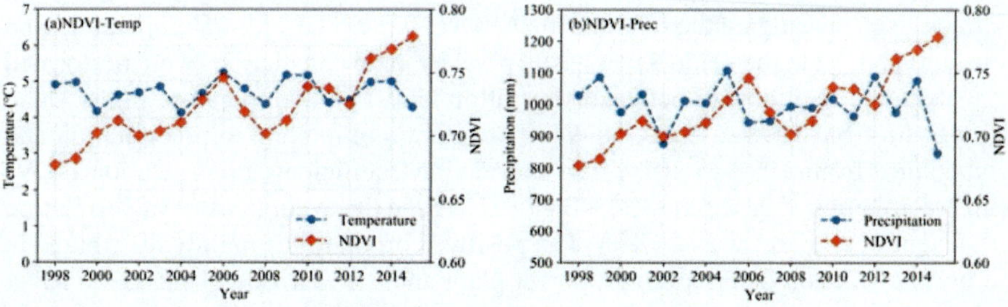

Fig. 2.4 Annual mean of NDVI and meteorological factors

The multi-year average precipitation is 997 mm, with the maximum annual average precipitation occurring in 2005 as 1108.03 mm and the minimum value occurring in 2015 as 845.71 mm. The coefficients of variation of annual mean precipitation and temperature are 6.79% and 7.21%, respectively, indicating that the annual mean precipitation and temperature of Dadu River basin have strong temporal stability.

The sliding correlation coefficient was used to check the stability of the correlation between NDVI and temperature and precipitation. The correlation coefficients between NDVI and temperature and precipitation were calculated for with a sliding window length of 9a, and the correlation coefficient series for each time period from 2002 to 2011 were obtained as shown in Fig. 2.5, according to which the stability characteristics of the correlation coefficients between NDVI and meteorological factors over time were analyzed. NDVI and temperature show a positive correlation with a mean correlation coefficient of 0.39, and the correlation shows an increasing trend from 2002 to 2004, and the correlation shows a more stable fluctuation from 2004 to 2010, and the correlation coefficient starts to decrease after 2010. The correlation coefficient between NDVI and precipitation is −0.021, and the correlation coefficient increases gradually from 2002 to 2006, and becomes positive in 2004 and negative in 2007, and then becomes positive again in 2010, with no stable positive or negative correlation. In general, the overall vegetation NDVI of the Dadu River basin was positively correlated with temperature, but there was no stable correlation with precipitation.

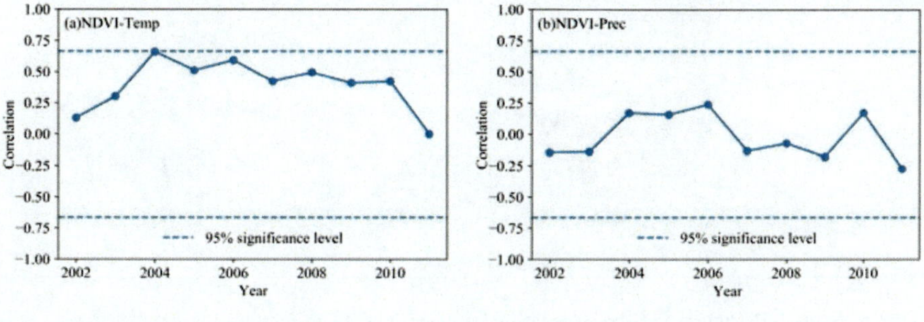

Fig. 2.5 Sliding correlation coefficient between NDVI and meteorological factors

In order to investigate the relationship between precipitation and temperature and vegetation NDVI in the Dadu River basin, pixel by pixel calculations were performed to analyze the relationship between vegetation NDVI and temperature in the Dadu River basin from 1998 to 2009. NDVI of vegetation in relation to precipitation and temperature from 1998 to 2019. The correlation coefficients of vegetation NDVI with precipitation (Fig. 2.6a) were −0.87 ~ 0.71, and the average value of correlation coefficient of all pixel was −0.103. The positive correlation is mainly distributed in the north of Shuangjiangkou hydropower plant and the east side of the Dadu River main stream from Luding hydropower plant to Shuangjiangkou hydropower plant, as well as the area around Gongga Mountain itself with less precipitation, occupying 33.96% of the total area; the negative correlation area is mostly concentrated in most of the area downstream of Luding hydropower plant, as well as the west of the Dadu River main stream and other areas with more annual precipitation, occupying 66.04% of the total area. The spatial distribution of the correlation between precipitation and NDVI illustrates that the contribution of precipitation to vegetation growth is not necessarily positive but two-sided, with too much precipitation limiting vegetation growth and the right amount of precipitation promoting vegetation growth.

The distribution of correlation coefficients between NDVI and temperature is shown in (Fig. 2.6b), and the average correlation coefficient of all pixel is 0.0727.

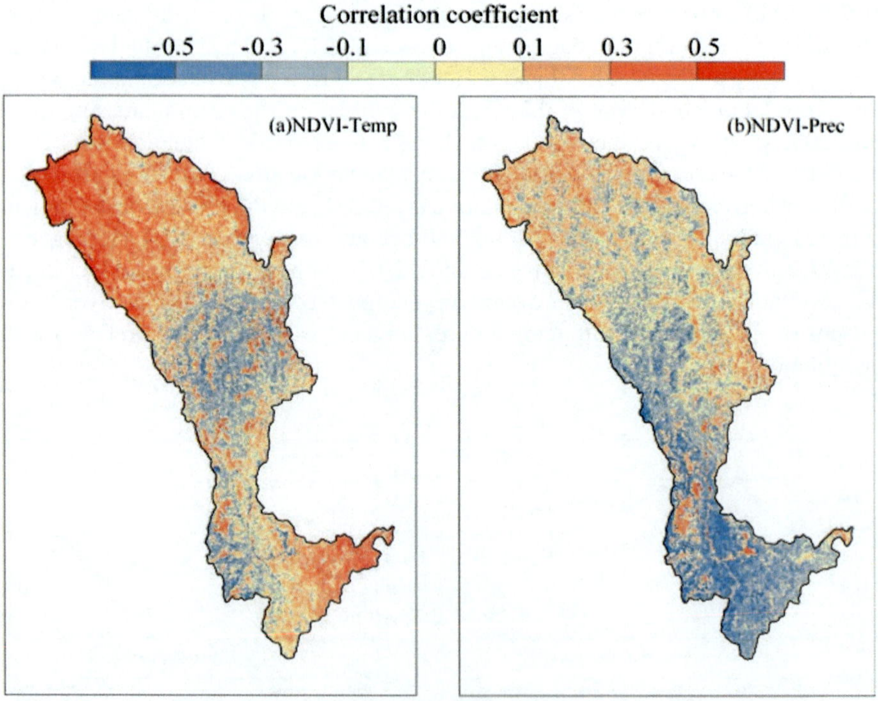

Fig. 2.6 Spatial distribution of the correlation between NDVI and meteorological factors

The positive correlation area is mainly distributed to the north of Shuangjiangkou hydropower plant and most of the area from Cascade Gorge hydropower plant to the downstream plain and the high altitude area within the area from Shuangjiangkou hydropower plant to Cascade Gorge hydropower plant, occupying 60.80% of the total basin area; the negative correlation area is mainly concentrated in most of the low altitude area from below Shuangjiangkou hydropower plant to Cascade Gorge hydropower plant, occupying 39.13% of the total area. Different from the correlation distribution of NDVI and precipitation, NDVI and temperature are still positively correlated in the plain areas where the annual average temperature is higher in the downstream, except for the area where the annual average temperature is lower in the upstream.

In conclusion, the correlation between temperature and precipitation and vegetation NDVI in the Dadu River basin is not completely positive but spatially heterogeneous. In areas with low temperature and precipitation, temperature and precipitation can promote vegetation growth, while excessive precipitation and temperature may lead to a decrease in vegetation NDVI. And in the area with abundant precipitation and low temperature like the Dadu River basin, temperature has a greater influence on NDVI than precipitation, especially for the higher elevation area in the northern part of the basin and the eastern part of the lower reaches, temperature has an obvious positive relationship on NDVI, and in the northern high elevation area, due to the original low precipitation and low temperature, temperature and precipitation together show a positive influence in this area.

a. **Characteristics of the influence of hydropower plant construction on regional NDVI changes**

In order to analyze the influence of hydropower plant construction on NDVI within the influence area of the plant, the four largest reservoir hydropower plant s in the main stream of the Dadu River were selected and the slope of NDVI change within the influence area of the plant and the distribution of multi-year NDVI before and after the start of construction were calculated for 22 years. temporal and spatial distribution variation characteristics of NDVI.

Subscript 1 represents the multi-year NDVI average before the construction of the hydropower plant, subscript 2 represents the multi-year NDVI average after the start of the construction of the hydropower plant, and subscript 3 represents the slope of NDVI variation during the study period. Figure 2.7($a_1 \sim d_1$) shows the multi-year NDVI mean values of each hydropower plant before construction. The largest NDVI mean value before construction is 0.794 at Dagangshan, the smallest is 0.736 at Cascade Gorge, 0.768 at Changhe Dam, and 0.747 at Monkey Rock. The river locations of Waterfall Gorge and Monkey Rock have relatively low NDVI values. Figure 2.7($a_2 \sim d_2$) show the average multi-year NDVI values after the construction of the hydropower plant. After the construction of the hydropower plant, the NDVI difference between the river area and the reservoir perimeter increases significantly compared with that before the construction, and the change of NDVI mean value of all the hydropower plant influence areas is positive, at this time, the largest NDVI mean value is 0.827 in Dagangshan, the smallest mean value is 0.764 in Cascade Gorge, 0.818 in Changhe Dam, and 0.816 in Monkey Rock, where the multi-year

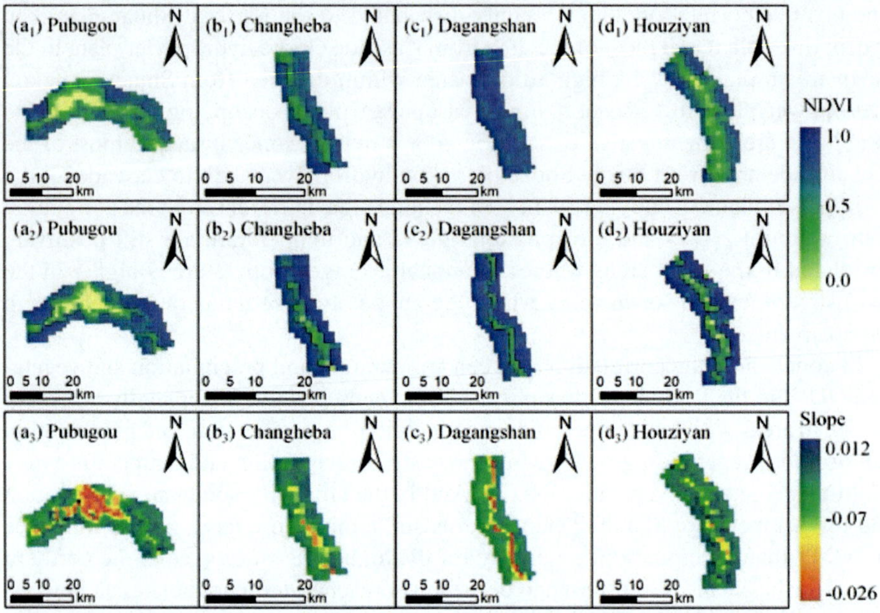

Fig. 2.7 Characteristics of NDVI changes in the area of influence of the plant

NDVI mean value of Monkey Rock increases The largest increase in NDVI was 0.0689 at Monkey Rock, and the least significant increase was 0.0282 at Waterfall Gorge. Figure 2.7($a_3 \sim d_3$) show the direction and magnitude of NDVI changes in the influence range. The area of Slope < 0 of each plant is only distributed at the river, and the rest of the area shows an increasing trend, among which the waterfall ditch plant is the plant with the largest inundation area therefore it is also the largest area of Slope < 0 among the four plants, accounting for 25.27% of the total area.

The rate of change of NDVI in the area of influence of each plant before and after the construction of the dam Slope is shown in Fig. 2.8. From the trend, the NDVI change rate in the influence area of all hydropower plant s after the construction of dams still showed a significant increasing trend, among which the Slope values after the construction of Monkey Rock and Changhe dams started were larger than before the construction of dams, and the NDVI change rate increased from 0.0037/a ($p < 0.05$) to 0.0116/a ($p < 0.05$) and from 0.0024/a ($p > 0.05$) to 0.0093/a ($p > 0.05$), respectively. to 0.0093/a ($p < 0.05$); while the Slope values after the construction of Cascade Gorge and Dagang Mountain started were smaller than before the dams were built, and the rate of change of NDVI decreased from 0.0061/a ($p > 0.05$) to 0.0014/a ($p > 0.05$) and from 0.0050/a ($p < 0.05$) to 0.0021/a ($p > 0.05$), respectively. In the influence area of the selected hydropower plant s except for the Dagangshan hydropower plant, where the NDVI was not affected in the pre-construction period and still maintained an increasing trend, the NDVI of the other three hydropower plant s decreased in the pre-construction period. Among them, the NDVI of the

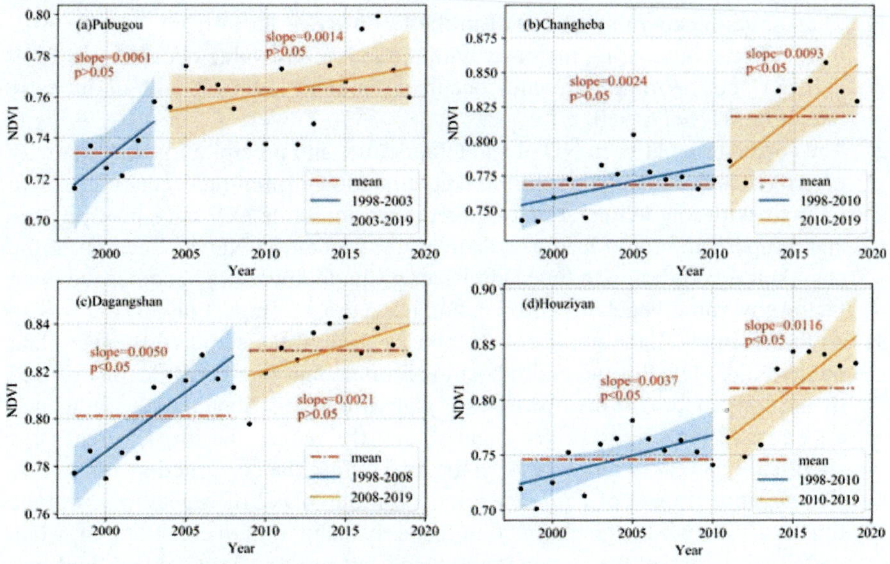

Fig. 2.8 NDVI change rate before and after dam construction

Cascade Ditch hydropower plant decreased continuously from 0.775 to 0.737 during the period from the cut-off in 2005 to the completion of the main project in 2011, and then recovered rapidly. Changheba hydropower plant had a small decrease in NDVI from 0.772 to 0.767 from the start of construction in 2010 to 2013, and then rapidly increased to 0.837 in 2014. Monkey Rock hydropower plant had a small decrease in NDVI from 2011 when it was cut off to 2012 and then recovered rapidly after 2013. The average NDVI in the impact area of Dagangshan hydropower plant did not change significantly in the first few years after the construction started, but started to decline continuously after all units were put into operation in 2015.

From the results of NDVI change in the influence area of the hydropower plant, the negative impact of dam construction on the NDVI of vegetation is mainly concentrated in the early stage of dam construction, which will cause a certain decrease in NDVI but the disturbance will recover quickly after the dam construction.

2.4 Conclusion

In this study, the NDVI change rate before and after dam construction was evaluated by using the synthetic data set based on SPOT/VEGETATION and the national meteorological interpolation data, selecting the Dadu River basin and the four hydropower plant s on the Dadu River main stream as the study area, and studying the NDVI change characteristics and influencing factors from 1998 to 2019. basin cascade hydropower development on vegetation cover, and the study concluded the following.

(1) The annual mean value of NDVI in the Dadu River basin from 1998 to 2019 maintained a fluctuating increase, with the maximum value occurring in 2019 at 0.781 and the minimum value occurring in 1998 at 0.677, with an increase rate of 0.00476/a ($p < 0.05$).
(2) The correlation between NDVI and temperature and precipitation was spatially heterogeneous, with temperature and precipitation promoting vegetation growth in areas with low temperature and precipitation, while excessive precipitation and temperature could lead to a decrease in vegetation NDVI. Because of the abundant precipitation in the Dadu River basin, temperature becomes the main factor governing vegetation growth and has a greater impact on NDVI.
(3) At the basin level, the impact of the construction of the step hydropower plant on the NDVI level of the entire basin is limited, and the NDVI decline caused by the construction of the hydropower plant is only distributed in the inundated area around the original river channel, and the rest of the impact area has a significant increase compared with the area before the construction.
(4) The negative impact of dam construction on the NDVI of vegetation from the time scale is mainly concentrated in the early stage of dam construction, when the NDVI may decline to a certain extent, but not to a large extent, and will not affect the overall trend, and the disturbance will recover quickly after the construction of the dam.

The spatial resolution of the data used in this study is 1 km, and the use of low resolution data for analysis may overlook some more detailed results. NDVI is now widely used but NDVI also suffers from the problems of easy saturation in areas with high vegetation cover and incomplete correction of atmospheric effects, about which the problem of easy saturation exists in many studies (Song et al. 2017). Further selection of optimized vegetation index is needed to improve the sensitivity to high biomass areas and enhance the monitoring of vegetation. Previous studies in many regions have found lags in the response of NDVI to changes in meteorological elements, and the use of annual average data in this study may not accurately reflect the correlation between the two. Changes in vegetation and habitat climate due to hydropower development need to be studied with more detailed peripatetic data and longer time scales.

Acknowledgements This research was the financially supported by Key R&D Projects of Sichuan Science and Technology Plan (2022YFG0120).

References

Botelho A, Ferreira P, Lima F, Costa Pinto L, Sousa S (2017) Assessment of the environmental impacts associated with hydropower. Renew Sustain Energy Rev 70:896–904
Chu H, Venevsky S, Wu C, Wang M (2019) NDVI-based vegetation dynamics and its response to climate changes at Amur-Heilongjiang River Basin from 1982 to 2015. Sci Total Environ 650:2051–2062

Hao A, Xue X, Duan H, Peng F, You Q (2023) Different spatiotemporal variations in seasonal NDVI and their climatic driving forces of a typical grassland on the Qinghai—Tibetan Plateau. Acta Ecol Sin 43(1):1–12

He B, Yan T, Huang Y, Long Z, Yan J, Zhao F, Chen Y, He Z, Du J (2021) Analysis of fish resources status in the upper reaches of the Dadu River (in Chinese). Freshwater Fisheries 51:38–45

Holben BN (1986) Characteristics of maximum-value composite images from temporal AVHRR data. Int J Remote Sens 7:1417–1434

Hutchinson MF (1995) Interpolating mean precipitation using thin plate smoothing splines. Int J Geograph Inf Syst 9:385–403

Li J, Dong S, Yang Z, Peng M, Liu S, Li X (2012) Effects of cascade hydropower dams on the structure and distribution of riparian and upland vegetation along the middle-lower Lancang-Mekong River. For Ecol Manage 284:251–259

Liu Y, Lu H, Tian P, Qiu L (2022a) Evaluating the effects of dams and meteorological variables on riparian vegetation NDVI in the Tibetan Plateau. Sci Total Environ 831:154933

Liu Y, Li Z, Chen Y, Li Y, Li H, Xia Q, Kayumba PM (2022b) Evaluation of consistency among three NDVI products applied to High Mountain Asia in 2000–2015. Remote Sens Environ 269:112821

Liu K, Du L, Hou J, Hu Y, Zhu YG, Gong F (2018) Spatiotemporal variations of NDVI in terrestrial ecosystems in China from 1982—2012. Acta Ecologica Sinica 38(1):1885–1896

Ma Y, Zuo L, Gao J, Liu Q, Liu L (2021) The karst NDVI correlation with climate and its BAS-BP prediction based on multiple factors. Ecol Ind 132:108254

Ouyang W, Hao FH, Zhao C, Lin C (2010) Vegetation response to 30years hydropower cascade exploitation in upper stream of Yellow River. Commun Nonlinear Sci Numer Simul 15:1928–1941

Peng W, Kuang T, Tao S (2019) Quantifying influences of natural factors on vegetation NDVI changes based on geographical detector in Sichuan, western China. J Clean Prod 233:353–367

Song W, Mu X, Ruan G, Gao Z, Li L, Yan G (2017) Estimating fractional vegetation cover and the vegetation index of bare soil and highly dense vegetation with a physically based method. Int J Appl Earth Obs Geoinf 58:168–176

Yang J, Dong J, Xiao X, Dai J, Wu C, Xia J, Zhao G, Zhao M, Li Z, Zhang Y, Ge Q (2019) Divergent shifts in peak photosynthesis timing of temperate and alpine grasslands in China. Remote Sens Environ 233:111395

Yi Y, Zhou Y, Song J, Zhang S, Cai Y, Wei Y, Yang Z (2019) The effects of cascade dam construction and operation on riparian vegetation. Adv Water Resour 131:103206

Zhang P, Cai Y, Yang W, Yi Y, Yang Z, Fu Q (2020) Contributions of climatic and anthropogenic drivers to vegetation dynamics indicated by NDVI in a large dam-reservoir-river system. J Clean Prod 256:120477

Chapter 3
Research on Installation Height Optimization Technology of Energy Dissipation Box of Trailing Suction Hopper Dredger

Hui Sun, Jiacheng Luo, Yuchi Hao, and Jiaming Qu

Abstract Aiming at the installation height optimization requirement of the key equipment for energy dissipation in the 15000 m^3 level trailing suction hopper dredger, which is the first self-developed powered by Liquefied Natural Gas clean energy in China, the influence of the installation height of energy dissipation box on loading efficiency was studied from the perspective of water and sediment movement by means of numerical simulation. It is found that for the target soil, as the energy dissipation box installation height decreases, the loading quality and natural soil quality increase, and the total spillage loss decreases. The results suggest that the optimum installation height of energy dissipation box is 9–10.5 m from the center of the box to the bottom of the hopper. The research can provide a theoretical basis for the design of key energy dissipation equipment of 15000 m^3 level LNG trailing suction hopper dredger, the improvement of the efficiency of the loading construction and the reduction of environmental pollution, and also provide technical support for the subsequent construction and transformation of the trailing suction hopper dredgers.

Keywords Trailing suction hopper dredger · Energy dissipation box · Loading quality · Natural soil quality · Spillage loss

3.1 Introduction

With the depletion of traditional fossil fuels and serious global emission problem, the world is increasingly demanding for environmental protection, which leads to the development trend of efficient, energy-saving and green shipping key technologies in response to the national call (Jinlong and Shimao 2016). Since the dredging construction is often carried out in the area close to the city such as the port channel, the pollutant emissions of the dredger directly affect the environmental quality of the

H. Sun · J. Luo · Y. Hao · J. Qu (✉)
CCCC National Engineering Research Center of Dredging Technology and Equipment Co. Ltd, Yangpu District, 28F, No. 1088 Yangshupu Road, Shanghai, China
e-mail: qujiaming@cccltd.cn

H. Xu (ed.), *Proceedings of the 5th International Symposium on Water Resource and Environmental Management*, Environmental Science and Engineering,
https://doi.org/10.1007/978-3-031-31289-2_3

construction area and the adjacent city (Xianli et al. 2016). Therefore, the research on the construction technology of Liquefied Natural Gas (LNG) clean energy power trailing suction hopper dredger (TSHD) is a technical guarantee for complying with the needs of national green development and promoting the green transformation of dredging equipment.

Aiming at the optimization design of the hopper system of TSHD, at present, domestic and foreign research institutions and fabrication companies mainly consider factors such as equipment material, structural strength, structural shape, structural size, wear performance, process difficulty and outflow mode (Xinxing et al. 2018; Guorui and Shumin 2012), but rarely from the perspective of construction efficiency after the ship put into use to optimize the layout of local key equipment. As a result, once the ship is delivered, the arrangement of key equipment such as energy dissipation box may become the upper limit of its construction efficiency. Existing researches on the layout optimization of energy dissipation key equipment mostly focus on the type, outlet direction and shape of energy dissipation box (Zhenlang and Zhan 2012; Minhua et al. 2021), but there is no mature research on the height layout.

Therefore, the research on the optimization technology of the installation height of the energy dissipation box is carried out by means of numerical simulation from the perspective of the loading construction process, and the influence on loading construction efficiency is analyzed, which can provide technical assistance for the optimization of the energy dissipation box layout of the first self-developed 15,000 m^3 level LNG TSHD in China.

3.2 Theoretical Basis of Simulation Model for Loading Construction

3.2.1 Fundamental Equations of Fluid

Assuming that the simulated fluid is Newtonian fluid which is incompressible and viscous, the fundamental equations of fluid motion are as follows.

The continuity equation is

$$V_F\frac{\partial\rho}{\partial t}+\frac{\partial}{\partial x}(\rho uA_x)+R\frac{\partial}{\partial y}\left(\rho vA_y\right)+\frac{\partial}{\partial z}(\rho wA_z)+\xi\frac{\rho uA_x}{x}=R_{DIF}+R_{SOR} \tag{3.1}$$

The momentum equation can be expressed as

$$\left\{\begin{array}{l}\frac{\partial u}{\partial t}+\frac{1}{V_F}\left(uA_x\frac{\partial u}{\partial x}+vA_y\frac{\partial u}{\partial y}+wA_z\frac{\partial u}{\partial z}\right)=-\frac{1}{\rho}\frac{\partial P}{\partial x}+G_x+f_x\\ \frac{\partial v}{\partial t}+\frac{1}{V_F}\left(uA_x\frac{\partial v}{\partial x}+vA_y\frac{\partial v}{\partial y}+wA_z\frac{\partial v}{\partial z}\right)=-\frac{1}{\rho}\frac{\partial P}{\partial y}+G_y+f_y\\ \frac{\partial w}{\partial t}+\frac{1}{V_F}\left(uA_x\frac{\partial w}{\partial x}+vA_y\frac{\partial w}{\partial y}+wA_z\frac{\partial w}{\partial z}\right)=-\frac{1}{\rho}\frac{\partial P}{\partial z}+G_z+f_z\end{array}\right\} \tag{3.2}$$

where V_F represents the fluid fractional volume, ρ represents the density of fluid, R_{DIF} and R_{SOR} represent the turbulent diffusion term and mass source, respectively. u, v and w represent different velocity components at coordinate directions. A_x, A_y and A_z represent the fractional area of fluid at coordinate directions. (G_x, G_y, G_z) and (f_x, f_y, f_z) represent body accelerations and viscous accelerations, respectively. P represents the fluid pressure.

3.2.2 *Fundamental Equations of Sediment*

For the loading process of TSHD, the simulaton mainly considers four kinds of sediment movement including suspension, advection, erosion and deposition.

For suspended sediment, it will undergo convection diffusion with fluid. The concentration calculation formula is as follows.

$$\frac{\partial C_{s,i}}{\partial t}+\nabla\cdot\left(C_{s,i}u_{s,i}\right)=\nabla\cdot\nabla\left(DC_{s,i}\right) \tag{3.3}$$

$$u_{s,i}=\overline{u}+u_{settle,i}c_{s,i} \tag{3.4}$$

$$c_{s,i}=\frac{C_{s,i}}{\rho_i} \tag{3.5}$$

where $C_{s,i}$ represents the mass concentration, $u_{s,i}$ represents the sediment velocity, D represents the diffusion coefficient, $\bar{u}$ represents the mixture mean velocity of fluid and sediment, $c_{s,i}$ represents the sediment volume concentration.

For sediment erosion, the empirical formula of the entrainment lift velocity (Mastbergen and Berg 2010) is used here.

$$u_{lift}=\alpha n_s d_*^{0.3}(\theta-\theta_{cr})^{1.5}\sqrt{\frac{\|g\|d_s\left(\rho_s-\rho_f\right)}{\rho_f}} \tag{3.6}$$

$$d_*=d_s\left[\frac{\rho_f\left(\rho_s-\rho_f\right)\|g\|}{\mu^2}\right]^{\frac{1}{3}} \tag{3.7}$$

where α represents the entrainment coefficient, n_s represents the outside normal direction of the packed sediment bed surface, θ represents the local Shields number, θ_{cr} represents the critical Shields parameter, g represents gravitational acceleration, μ represents the fluid viscosity, d_s represents the sediment particle size, ρ_s represents the sediment density, while ρ_f represents the fluid density.

The local Shields number θ can be calculated by

$$\theta = \frac{\tau}{\|g\| d_s (\rho_s - \rho_f)} \tag{3.8}$$

where τ represents the local shear stress on sediment.

It is very important for sediment incipience and scour of critical Shields parameter θ_{cr}, which can be calculated by the Shields-Rouse equation (Soulsby 1998).

$$\theta_{cr} = \frac{0.3}{1 + 1.2d_*} + 0.055\left[1 - \exp(-0.02d_*)\right] \tag{3.9}$$

where d_* represents the dimensionless sediment diameter, $d_* = d_{50}\left[\frac{\rho_f(\rho_s - \rho_f)\|g\|}{\mu^2}\right]^{\frac{1}{3}}$.

On a sloped bed, the critical Shields number needs to be corrected by the sediment repose angle.

$$\theta_{cr}\prime = \theta_{cr} \frac{\cos\psi \sin\kappa + \sqrt{\cos^2\kappa \tan^2\phi - \sin^2\psi \sin^2\kappa}}{\tan\phi} \tag{3.10}$$

where κ represents the angle between the vertical line of deposition surface and gravitational acceleration g, φ represents the repose angle of sediment, while ψ represents the angle between the slope direction and the flow direction of the fluid.

For bed-load transport sediment, the empirical formula (Meyer and Müller 1948) is written as

$$\Phi = \beta(\theta - \theta\prime_{cr})^{1.5} \tag{3.11}$$

where β represents the transport coefficient, Φ represents the non-dimensional bed load transfer rate. The relationship between it and the single-width transport rate q is as follows

$$q = \Phi\left[\|g\|\left(\frac{\rho_s - \rho_f}{\rho_f}\right)d_s^3\right]^{\frac{1}{2}} \tag{3.12}$$

The bed load sediment thickness δ (Van 1984) can be expressed as

$$\frac{\delta}{d_s} = 0.3d_*^{0.7}\left(\frac{\theta}{\theta\prime_{\mathrm{cr}}} - 1\right)^{0.5} \tag{3.13}$$

Therefore, the sediment velocity of bed load transport q_b is shown as

$$u_b = \frac{q_b}{\delta f_b} \tag{3.14}$$

3.3 Model Validation of Loading Construction Simulation

3.3.1 Verification of Flow Field

In order to validate the feasibility of the loading construction simulation in the flow field, taking the physical hopper model in laboratory as the prototype, the arrangement of physical hopper model with five measuring points and the three dimensional numerical simulation of the same scale are shown in Fig. 3.1. The experimental data and simulated data of the flow velocity in different directions of the five measuring points were compared and analyzed as shown in Fig. 3.2.

The results show that the simulation results of the longitudinal velocity, transverse velocity and vertical velocity of each measuring point on different water depth planes are consistent with the experimental data, and the error is small, which indicates that the numerical model has good accuracy in the simulation of flow field.

3.3.2 Verification of Loading and Overflow Characteristics

To verify the feasibility of the numerical model in the simulation of the loading and overflow characteristics, the three dimensional hopper model of the Xinhaihu 4 TSHD was established based on the design information, as shown in Fig. 3.3.

According to the dredging construction data of Xinhaihu 4 in the Yangtze Estuary, the measured loading data of one ship which can represent the average construction level was selected for simulation verification. The mainly input conditions are shown in Table 3.1.

The three dimensional simulation of the overflow process is shown in Fig. 3.4, and the comparison between the simulation and the measured construction data represented by the loading quality and the natural soil quality is shown in Fig. 3.5. The result shows that from the beginning of loading to overflow and the subsequent overflow process, the simulation results of loading quality and the natural soil quality are basically consistent with the measured data.

Furthermore, compared with the measured data of Xinhaihu 4 and simulation data at the end of loading construction time, the errors of simulated loading quality and the natural soil quality are 5.53% and 7.96%, respectively, indicating that the loading and

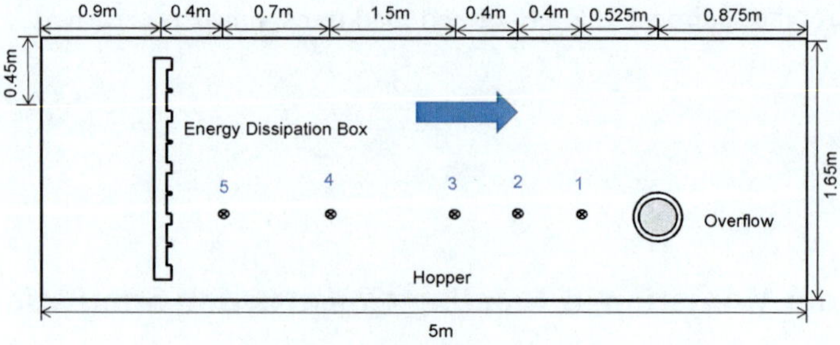

(1) Arrangement of physical hopper model and five measuring points

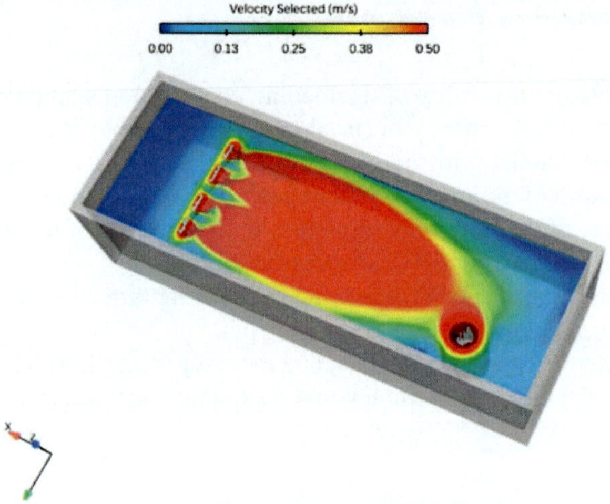

(2) numerical simulation of flow field in hopper

Fig. 3.1 Physical model arrangement and numerical simulation of flow field in hopper

overflow characteristics can be simulated well by the loading construction numerical model of Xinhaihu 4 for the soil in the Yangtze River estuary.

3.4 Height Optimization of Energy Dissipation Box for LNG TSHD

3.4.1 Three Dimensional Hopper Modeling

The three dimensional hopper model of the 15000 m^3 level LNG TSHD was established based on the design information, as shown in Fig. 3.6.

Fig. 3.2 Comparison between simulated and experimental results of flow velocity at different positions

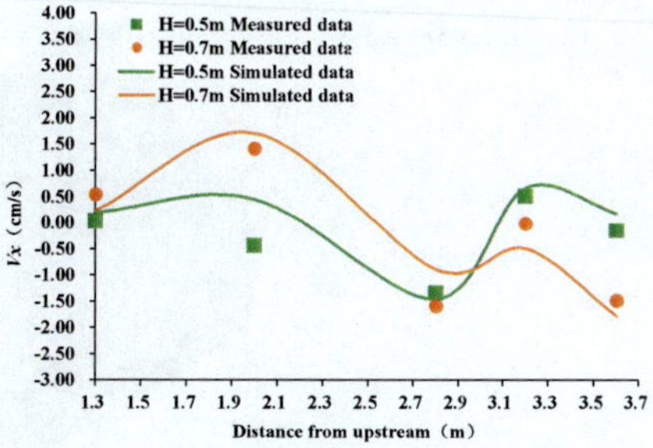

（1）Longitudinal velocity V_x

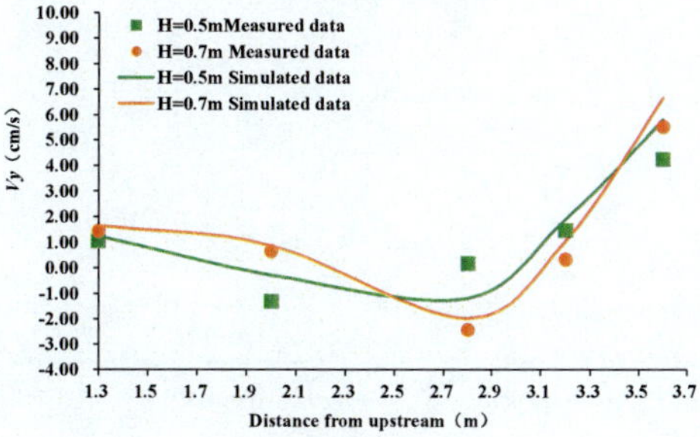

（2）Transverse velocity V_y

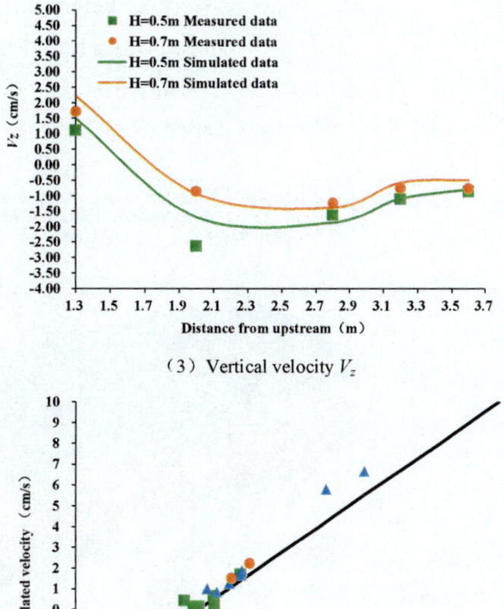

（3）Vertical velocity V_z

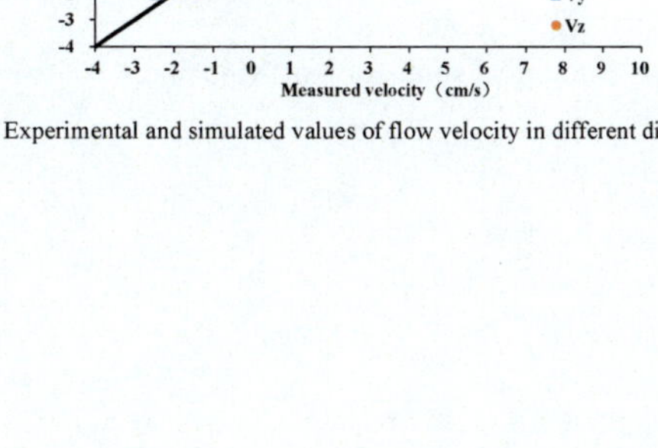

（4）Experimental and simulated values of flow velocity in different directions

Fig. 3.3 Three dimensional model of Xinhaihu 4 hopper

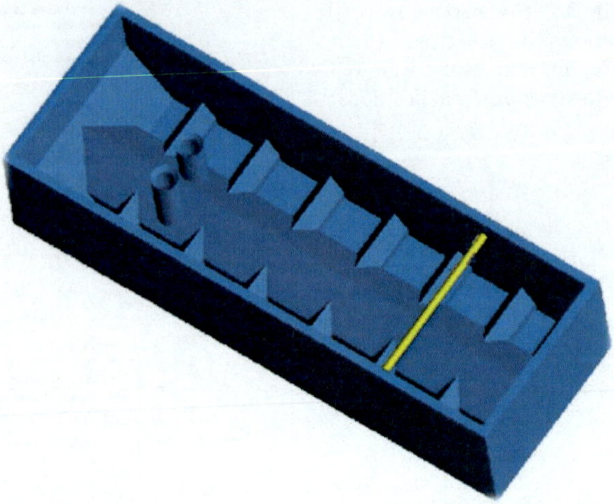

Table 3.1 Mainly input conditions of loading simulation of Xinhaihu 4

Parameter	Value
Inflow flow rate [m^3/s]	6.94
Inflow density [kg/m^3]	1160
Overflow height [m]	12.86
Initial water level [m]	0.4
Construction time [min]	114
Medium diameter of sediment [mm]	0.08

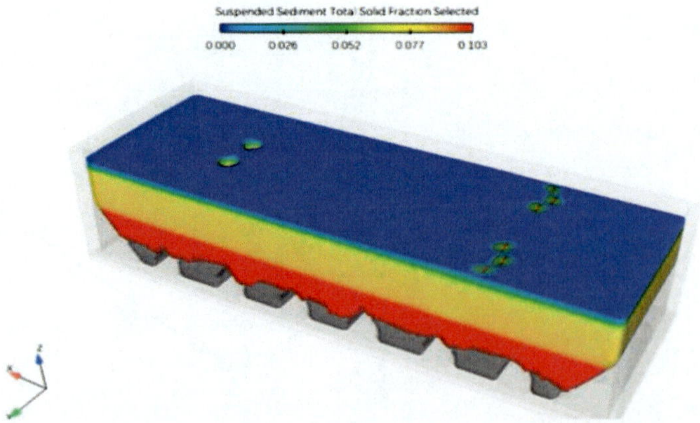

Fig. 3.4 Numerical simulation of Xinhaihu 4 construction

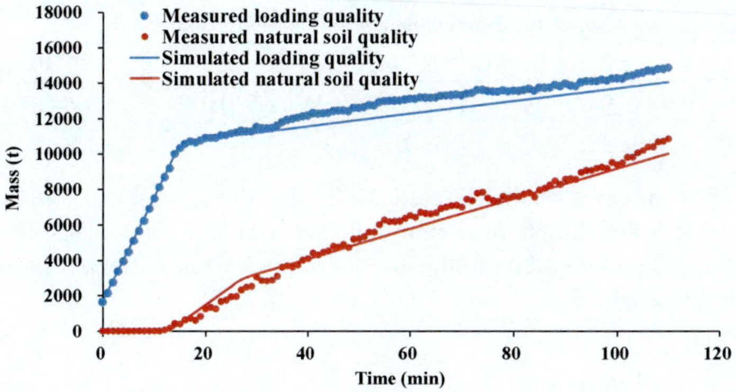

Fig. 3.5 Measured and simulated variations of loading quality and natural soil quality with time

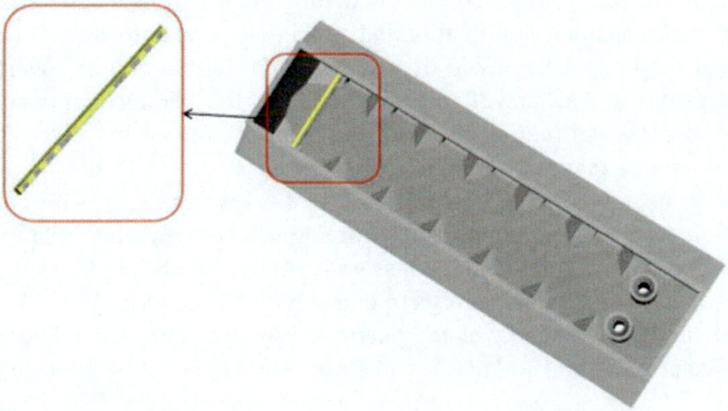

Fig. 3.6 Three dimensional model of the 15000 m^3 level LNG TSHD

3.4.2 Simulation Conditions and Design Cases

Based on the design information of the 15000 m^3 level LNG TSHD, the mainly input conditions for loading simulation were set as Table 3.2.

Table 3.2 Mainly input conditions for loading simulation of the 15000 m^3 level LNG TSHD

Parameter	Value
Inflow flow rate [m^3/h]	25,000
Inflow density [kg/m^3]	1170
Overflow height [m]	12.00
Initial water level [m]	0.7
Construction time [min]	90
Medium diameter of sediment [mm]	0.08

Table 3.3 Simulation cases of the different height for the energy dissipation box

No	1	2	3	4	5	6	7	8	9	10	11
H	0.90	0.86	0.82	0.78	0.75	0.71	0.67	0.63	0.60	0.56	0.52

The simulation cases of the different height for the energy dissipation box are shown in Table 3.3 to explore the height influence (the ratio of distance from center point of energy dissipation box to the bottom of the hopper to hopper height H) on the loading efficiency.

3.4.3 Results and Analysis

The simulation results, composed of key loading curves from 5 cases represented by loading quality, natural soil quality and spillage loss rate are shown in Fig. 3.7. It shows that the hopper begins to overflow around 30 min and the overflow rate gradually flattens out at around 80 min. Further more, the loading quality and natural soil quality increase with the decrease of the height of the energy dissipation box, and the spillage loss rate decreases.

In order to quantify the height influence of the energy dissipation box on the loading efficiency, the simulation results of relative loading quality, relative natural soil quality and relative total spillage loss with relative height of energy dissipation box for case of $H = 0.9$ at 90 min were compared. As shown in Fig. 3.8, with the decrease of the relative height of the energy dissipation box, the relative loading quality and relative natural soil quality increase, while the relative total spillage loss decreases.

According to the extent of change, three areas were divided, namely salient region, transition region and smooth-out region. In the salient region, as the height of the energy dissipation box decreases, the growth rate of loading quality and natural soil quality is the largest, so as to the reduction rate of total spillage loss. In the transition region, the change range slower, which changes gently in the smooth-out transition.

Therefore, the optimal height range of the energy dissipation box for the 15000m^3 level LNG TSHD is recommended 9–10.5 m in the transition region comprehensively considering the hopper structure layout and the influence of the energy dissipation box height on loading efficiency for the soil of the Yangtze Estuary.

3.5 Conclusions

The three-dimensional loading numerical model for TSHD proposed in this study is of good accuracy in the simulation of flow field and loading-overflow characteristics. Compared with the experimental data, the model accurately reflects the

Fig.3.7 Variation of loading quality, natural soil quality and spillage loss rate with time

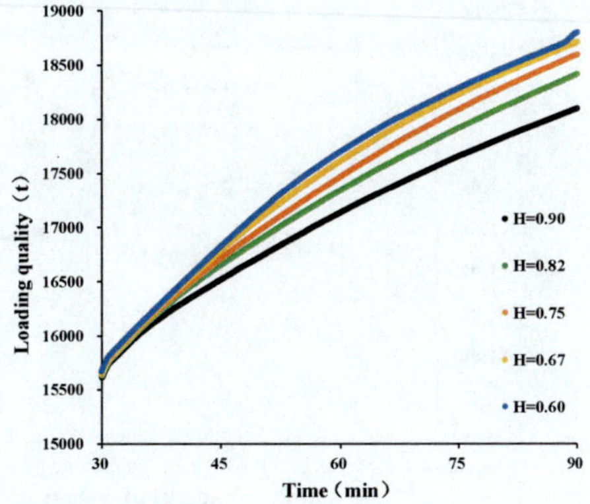

(1) Loading quality

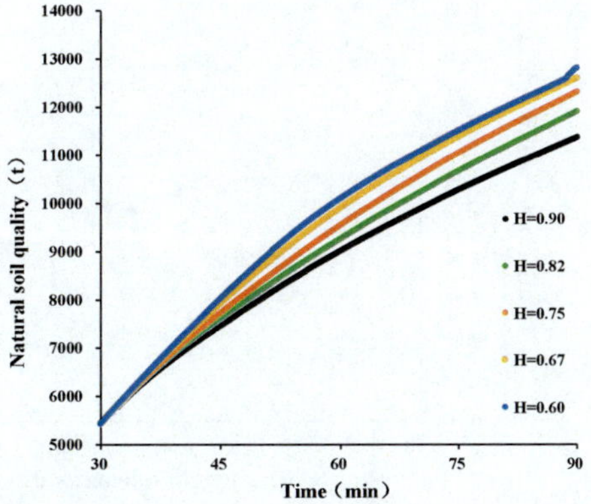

(2) Natural soil quality

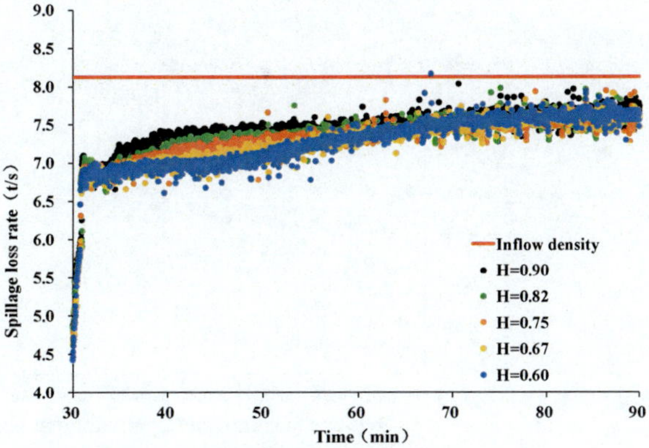

(3) Spillage loss rate

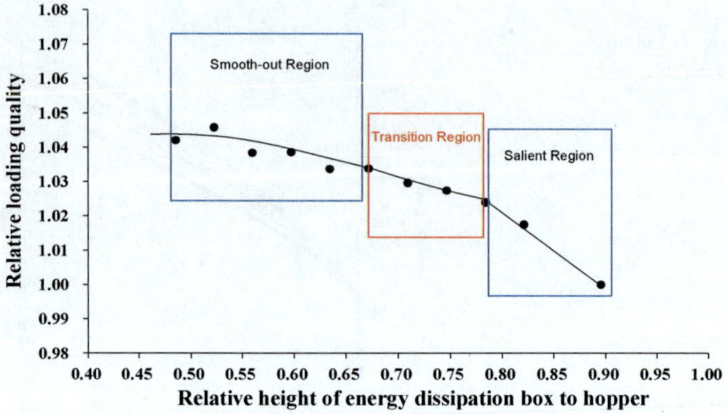

(1) Relative loading quality

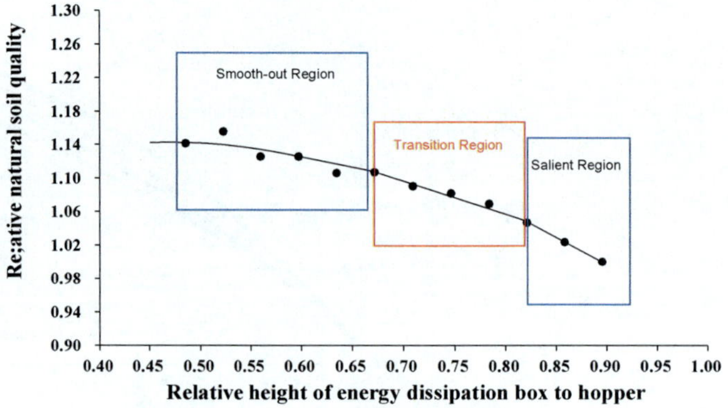

(2) Relative natural soil quality

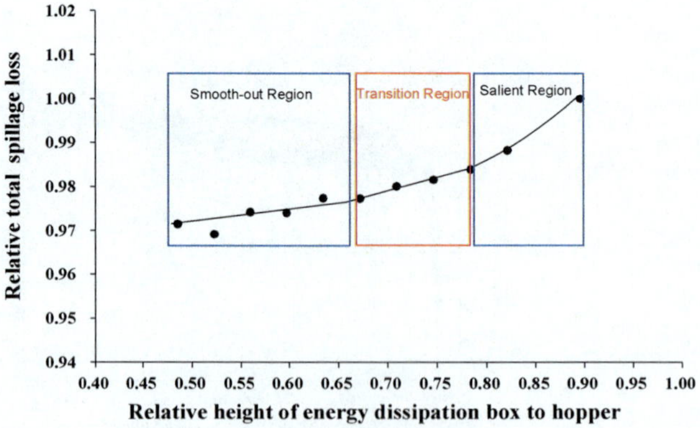

(3) Relative total spillage loss

Fig. 3.8 Variation of relative loading quality, relative natural soil quality and relative total spillage loss with relative height of energy dissipation box to hopper variation characteristics of flow field in the hopper. The simulation error of key loading parameters for Xinhaihu 4 construction in the Yangtze River estuary is within 8%.

For the first self-developed LNG-powered 15,000 m^3 level TSHD in China, with the decrease of the energy dissipation box installation height, the loading quality and natural soil quality increase, and the total spillage loss decreases for the Yangtze Estuary sediment.

It is recommended that the optimal installation height of the fixed energy dissipation box is within the range of 9–10.5 m. Under allowable condition, dynamically promoting the height of the energy dissipation box during the loading construction process is preferred.

Acknowledgements This work was supported by the National Key Research and Development Program of China (No. 2021YFB2601100).

References

Guorui Q, Shumin Z (2012) Structure design for mega trailing suction hopper dredger. Ship Engineering 36(2):1–8

Jinlong Z, Shimao Z (2016) Characteristic analysis of power unit for LNG-fueled trailing suction hopper dredger and discussion on selection of engine. China Harbour Eng 36(9):69–72

Mastbergen DR, Berg J (2010) Breaching in fine sands and the generation of sustained turbidity currents in submarine canyons. Sedimentology 50(4):625–637

Meyer PE, Müller R (1948) Formulas for bed-load transport. In: 2nd Proceeding of Congress IAHR. Stockholm, Sweden, pp. 39–64

Minhua S, Jing Y, Chen Z, Xiaodan M, Guojun H, Ke C (2021) Numerical study on energy dissipation structure characteristics of hopper of trailing suction hopper dredger. Port Waterway Eng 8:207–212

Soulsby R (1998) Dynamics of Marine Sands: a Manual for Practical Applications. Thomas Telford, London

Van RLC (1984) Sediment transport, Part I: bed load transport. J Hydraul Eng 110(10):1431–1456

Xianli W, Zaihong Y, Yingjie Z (2016) Application of dual-fuel engine on trailing suction hopper dredgers. Ship&boat 27(6):50–55

Xinxing L, Jianfeng W, Shaojun C (2018) Discussion on hopper grating design and overflow loss of large TSHD. China Harbour Eng 38(10):65–69

Zhenlang W, Zhan Z (2012) Study on the design of dredging system for large trailing suction dredge. Naval Architect Ocean Eng 2012(1):38–45

Chapter 4
Application Research of Educational Management in Fluidized Soil Engineering

Yuan Fang, Xinmiao Shi, Haoqing Xu, Pengming Jiang, Yiyan Lv, and Kaidong Zhang

Abstract A considerable quantity of the garbage, such as slag, steel slag, and so on, is generated during the development of cities. Furthermore, on-site garbage management is a significant concern. Currently, no research in the United States or elsewhere has directly combined slag and steel slag to create fluidized scrapped soil for on-site pit sidewall backfill, but it has been discovered that steel slag has a comparable mineral phase structure to cement and may be utilized as an alternate mix to cement. As a result, this paper investigates the engineering characteristic of fluidized soil for backfilling using scrapped soil, steel slag, and cement at various steel slag substitution rates (steel slag quality/cement quality + steel slag quality) and water content (water quality/cement quality + steel slag quality + scrapped soil quality). The findings reveal that increasing both the slag replacement rate and the water content can increase the flowability of the fluidized scrapped soil, with the water content having a significantly bigger impact on the flowability than the slag substitution rate.

Keywords Fluidized scrapped soil · Steel slag · Fluidity · Water content · Fill performance

4.1 Introduction

China's cement production is expected to reach 2.377 billion tons in 2020. Even if the pandemic is severe and the country goes into hibernation, the cement sector

Y. Fang
Physical Education College, Jiangsu University of Science and Technology, Zhenjiang 212100, China

X. Shi · H. Xu (✉) · P. Jiang · K. Zhang
Jiangsu Province Engineering Research Center of Geoenvironmental Disaster Prevention and Remediation of School of Architecture and Civil Engineering, Jiangsu University of Science and Technology, Zhenjiang 212100, China
e-mail: hank1nxu@just.edu.cn

Y. Lv
Zhejiang Huadong Construction Engineering Co. LTD, Hangzhou 310014, China

H. Xu (ed.), *Proceedings of the 5th International Symposium on Water Resource and Environmental Management*, Environmental Science and Engineering,
https://doi.org/10.1007/978-3-031-31289-2_4

can still achieve a 2.5% year-on-year rise. Furthermore, the carbon dioxide emitted by cement manufacturers accounts for approximately 8% of total carbon emissions emitted by human activities (Li et al. 2022; Luo et al. 2022). The use of cement, steel, and other building materials is rising along with China's urbanization process. As a result, some of the more common solid wastes produced during the urbanization process include slag and steel slag. A survey indicates that China's urban construction waste has accumulated to 10 billion tons and is continually increasing at a rate of 300 million tons annually. In addition, China produces approximately 100 million tons of steel slag annually, making it a significant producer of steel (Zhao and Zhang 2020). However, our slag and steel slag are now underutilized, and if they were to be stacked up indiscriminately, they would deplete available land resources. Additionally, some of the chemicals contained in them are simple to volatilize and move, damaging the water and soil around them and placing a significant strain on the ecological environment. However, it has been discovered that Portland cement clinker and steel slag share a similar mineral structure. As a result, it can be viewed as an alternative to making backfill material for engineering construction by mixing cement and slag. The two main issues with solid waste use may be effectively resolved and the resource exploitation of solid waste can be made possible if the mixture of slag and steel slag can be used in the field of pit sidewall backfill.

The geography and surrounding environment of the construction site frequently place restrictions on pit backfilling projects in China. For instance, it can be challenging to backfill in layers and compact in layers (American Concrete Institute 1999), especially when backfilling the pit sidewalls, because certain deep pit backfilling projects have limited working surfaces that prevent the use of heavy mechanical equipment. Therefore, a new type of backfill material is urgently needed that can solve the problems of narrow backfill space in the sidewalls of the pit, large backfill depths, and the existence of shaped cross-sectional structures in the backfill area. Controlled Low Strength Material (CLSM), as defined by the American Concrete Institute (ACI), is a self-compacting cementitious material that can be used as backfill and as an alternative to compacted fill in what may be termed fluidized soils. What's more, the use of fluidized soil is a good solution to the problem when backfilling the sidewalls of the foundation pit. However, most of our foundation pits are backfilled with high-quality materials such as clay, coarse sand, and stone dust slag, which are a waste of material for backfilling the sidewalls.

Predecessors have done more research on the resource utilization of slag and steel slag and fluidized soil. Yu et al. (2019) used sodium-based bentonite slurry to improve the slag soil, when bentonite accounted for the higher proportion of slurry, the lower the permeability coefficient of sandy soil. Wang (2015) improved the slag by adding four additives such as water, bentonite, foam, and polymer, which can reduce the wear of the slag on the shield and have a better lubrication effect. Using excavated spoil from the construction site and controlling the cement-to-water ratio and water-to-solids ratio to 0.5–0.7 and 0.7 respectively, Gemperline and Durham (2012) was able to produce a fluidized soil suitable for structural backfill. Wu and Lee (2011) not only fixed the cement-to-water ratio and the water-to-solids ratio at around 0.7 and 0.35 but also controlled the sand content of the sand to 40% to prepare a flowable soil suitable for bridges. At the same time, the unit weight and flowability of the produced flowable soil met the requirements. Huang et al. (2021) found that with appropriate control measures, steel slag can be used as an effective cement replacement aggregate in cement-stabilized base layers. Behiry (2013) used steel slag to improve pavement sub-base density, strength, and damage resistance with limestone aggregates. Achtemichuk et al. (2009) used recycled concrete aggregates such as slag and high calcium fly ash to produce composites that are flowable fillers, which can be used as composites requiring structural support and rapid hardening. Sheen et al. (2008) used oxidized slag and reduced slag instead of fine aggregates and cement to produce flowable soils with good fluidity at slumps greater than 20 cm. Wang and Chen (2016) used steel slag to replace fine aggregates to prepare flowable soils and found that the increase in slag replacement rate was detrimental to the flowability of flowable soils, whose slump decreased with the increase in slag replacement rate and the water secretion rate increased with the increase in slag replacement rate. Menina et al. (2016) found that treated oil sands waste can be used as an alternative material to sand in fluidized soils and that the use of oil sands waste increases the fluidity of fluidized soils. Nataraja and Nalanda (2008) found that all three materials, fly ash, rice husk ash, and quarry dust, can be used as mobile soil materials. Horiguchi et al. (2011) used incinerated sludge to replace fly ash in fluidized soils, adding incineration sludge would increase the water secretion rate. However, the water secretion rate of mobile soil with crushed stone powder instead of fine aggregate was lower than that of mobile soil with ordinary fine aggregate.

At present, there are relatively few studies on mobile soil at home and abroad, and no research has been carried out in some specific fields. Therefore, in this study, slag + steel slag + cement is used to make fluidized slag soil that not only meets the construction requirements but also uses steel slag to replace some of the cement to reduce the amount of cement used, at the same time, reduce the construction cost. The study was carried out to investigate the flowability, density, and other construction properties at different steel slag replacement rates and water contents.

4.2 Test Materials and Methods

4.2.1 Test Materials

The scrapped soil used in the test was taken from the sandy soil of Hongxin Square in Zhenjiang. After the residual soil was retrieved, it was first dried, crushed, and sieved in three steps. Then the basic physical properties of the soil samples were measured according to the Geotechnical Test Method Standard (GB/T 50,123–1999), as shown in Table 4.1.

Considering the production of the sample, a particle size less than 2 mm was used in the test, and the particle size distribution curve of test fluidized scrapped soil

Table 4.1 Physical properties of scrapped soil

Natural density	Specific gravity	Water content	Maximum dry density	Optimal water content
1.71 g/cm^3	2.69	12.8%	2.07 g/cm^3	20.94%

was obtained by Mastersizer 3000 laser particle size analyzer produced by Malvern Instruments Ltd. in the UK, and the results are shown in Fig. 4.1. The calculation can obtain that the residual slag $C_u = 11.59$, $C_c = 1.63$, and the gradation is good.

The steel slag used in the test comes from a steel mill in Zhengzhou, Henan, whose particle size gradation is shown in Fig. 4.1, considering the sample production and specification requirements. The steel slag selected is required to have a particle size less than 1/10 of the sample diameter, and take steel slag with a particle size of less than 4 mm. Before the test, the chemical composition of the steel slag was analyzed by an X-ray fluorescence spectrometer produced by PANalytical in the Netherlands. This is shown in Table 4.2.

Mason (1944) evaluated the activity of steel slag according to the alkalinity of steel slag. The alkalinity calculation is shown in Eq. (4.1,) and the relationship between the alkalinity and mineral composition of steel slag is shown in Table 4.3. The alkalinity of steel slag used in this study is 1.68, which belongs to dicalcium silicate slag, and

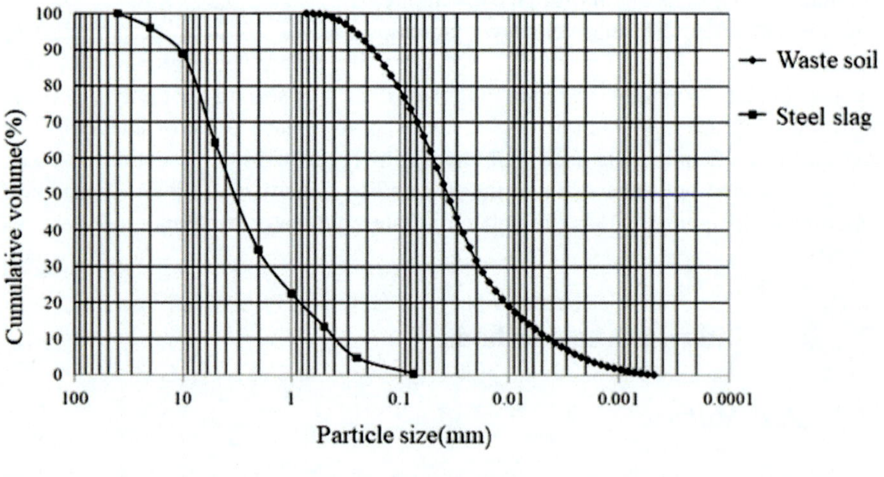

Fig. 4.1 Particle size distribution curve of test fluidized scrapped soil and original steel slag

Table 4.2 Test the chemical composition of the material

Material name	Main chemical composition										
	CaO	Fe_2O_3	SiO_2	Al_2O_3	MgO	SO_3	MnO	Cr_2O_3	Na_2O	P_2O_5	TiO_2
Steel slag	34.30	26.22	19.64	7.72	4.14	0.493	2.49	2.08	1.06	0.763	0.523
Cement	61.39	2.46	20.92	6.41	3.01	3.41	–	–	–	–	–

Table 4.3 Relationship between basicity and mineral composition of steel slag

Steel slag alkalinity	Types of steel slag	Main minerals
0.9~1.4	Peridot slag	Peridot, magnesium rhodonite, RO phase
1.4~1.6	Magnesium rose slag	Magnesium rhodonite, dicalcium silicate, RO phase
1.6~2.4	Dicalcium silicate residue	Dicalcium silicate, dicalcium ferrite, RO phase
>2.4	Tricalcium silicate residue	Tricalcium silicate, dicalcium silicate, dicalcium ferrite, RO phase

the main minerals are dicalcium silicate, dicalcium ferrate, and RO phase. Since the greater the alkalinity of the slag, the greater its activity (Cheng et al. 2009; Mo et al. 2019; Lu et al. 2018), the slag used for the study in this paper has a low activity.

$$P = CaO/(SiO_2 + P_2O_5) \tag{4.1}$$

Considering that the fluidized scrapped soil is to replace the compacted soil for backfilling, so it does not need too high strength. Therefore, the cement in this study used 32.5 composite silicate cement from a cement factory in Hanzhong. The results of the chemical composition of the cement were analyzed by X-ray fluorescence spectrometer as shown in Table 4.2.

4.2.2 Experimental Methods

(1) Fluidity

In this study, an acrylic plate with a length and width of 40 cm × 40 cm and a thickness of 5 cm was used according to the ASTMD-6103 specification. A scale was inscribed in the middle of the plate in all four directions and a minimum scale of 1 mm. A flow rate cylinder with an inner diameter of 80 mm, a height of 80 mm, and a wall thickness of 3 mm was used, and this cylinder had no bottom and no cover. The test method is: first wipe the acrylic plate and the flow cylinder clean, and then place the acrylic plate on the flat, horizontally, and free from vibration or other interference. At the same time, calm the flow of cylinder water to the center of the acrylic plate. After that, the mixture of well-stirred flowing soil was slowly poured into the flowing degree cylinder, so that it was slightly higher than the cylinder mouth, and then scraped with a scraper to make the upper surface of the specimen flush with the upper end of the flowing degree cylinder. Finally, the flowing degree cylinder was slowly raised along the vertical direction to let the flowing soil spread freely on the acrylic plate, and after the flowing soil, the mixture was stabilized. The diffusion distance in four directions was read out according to the scale on the acrylic plate, half of the sum of the four readings as the flow degree of this fluidized soil.

(2) Timed test

In this test, the flowability of the fluidized soil mixture was measured at five-time points after the completion of preparation: 0 min, 30 min, 60 min, 90 min, and 120 min. The instruments used in the test are the same as the flow test. The method is as follows: the initial flow value of the liquid soil mixture is measured according to the flow degree test, that is, the flow degree of 0 min. After completion, quickly scrape the flowing soil mixture into a bowl with a scraper. Try to clean it as much as possible to avoid the loss of materials during the scraping process. And then to prevent the water from evaporating too quickly, cover the bowl with plastic wrap. Then let it stand, wait until 30 min later, remove the plastic wrap, and stir the flowing soil mixture for 1 ~ 2 min. Finally, measure again according to the flow degree test until the flow degree of all time nodes is measured.

(3) Moisture retention

According to Ling et al. (2018), the bleeding rate of the general mixture is less than 5%, which is considered to be stable and has good water retention performance. Because there is no clear domestic specification to determine the bleeding rate of flowing soil, this test refers to the test method of testing the secretion rate of concrete in JTGE30-2005 Experimental Procedure for Cement and Cement Concrete for Highway Engineering, by: ① first weighing the mass of the bleeding rate cylinder with an electronic scale, recorded as G1, then pouring 1 L of well-mixed flowing soil mixture into the bleeding rate cylinder. During the pouring process, you can lightly tap the outer wall of the bleeding rate cylinder without vibrating and compacting. And then immediately weigh the mass of the bleeding rate cylinder at this time and record it as G2. Cover the top of the water secretion rate cylinder with cling film after completion, which is used as the starting point of timing at this time. ② After 2 h, remove the cling film from the top of the bleeding rate cylinder, then tilt it at a certain angle and place a pad underneath to facilitate water absorption from the cylinder. After that, place the water lurking in the tube into the measuring cylinder and record it as V, accurate to 1 mL.

(4) Filling performance

According to Lachemi et al. (2010), it was found that the filling rate of fluidized soil between 80 and 100% was considered to have good filling performance. Since there is no clear specification to determine the filling performance of flowable soil in China, this test is based on the actual engineering situation and the research methods of foreign scholars to design a model box as shown in Fig. 4.2 by ① estimating the mass of fluidized soil required to backfill the model box and mixing 10% more mass on top of that to ensure that the fluidized scrapped soil made and mixed is sufficient to backfill the model box. ② Take three density cylinders, pour the mixed fluidized scrapped soil into the cylinders,

Fig. 4.2 Foundation pit side wall model box

and calculate the density of each cylinder, when the error between the three is less than 15%, the average value of the three densities as the density of the backfilled fluidized scrapped soil. ③ Put the model box on the electronic scale to peel, then pour the weighed fluidized scrapped soil into the model box slowly. After the backfilling is completed, use the scraper to scrape the upper surface of the model box, and then weigh the model box immediately, and the backfilling is completed. scraped flat, and then immediately weighed. ④ Statistical data and calculate the filling rate, the filling rate is calculated.

4.2.3 Preparation and Test Scheme of Fluidized Residue

Before the test, first, according to the ratio of each group will be the amount of cement, steel slag, slag, and water that was weighed. Then weigh the good slag, cement, and steel slag to gradually add the mixing container, adding cement and steel slag should be as uniform as possible to cover the top of the slag. And then slowly add water and start stirring, until the water is all added. This process stirring strength should be careful not to be too large, otherwise not be mixed evenly with the material easily spilled out of the container. And then stirred for 3–5 min after the end of adding the material, and stopped stirring when the mixture of fluidized scrapped soil was mixed uniformly.

4.3 Results and Analysis

4.3.1 *Fluidity of Fluidized Scrapped Soil Under Different Water Content*

The fluidity of the fluidized scrapped soil can be well controlled by changing the water content (Thunder 2017). Its growth rate is relatively slow in the stage of water content of 58~2%, and faster in the stage of 62~64%. When the steel slag substitution rate is 50%, the flow value of the fluidized scrapped soil with a water content of 58~60% increases by 9.0%, and the flow value increases by 17.8% in the stage of 60% ~ 62% water content. The change of the flow degree of the fluidized scrapped soil with the water content can be divided into two stages, the first stage: with the increase of the water content, the growth trend of the flow degree is relatively slow (58~62% water content). At this time, the liquid soil presents a semi-solid state. As shown in Fig. 4.3a–c), there is no obvious water secretion on the surface of the sample, which is relatively good. The second stage: with the increase of water content, the growth trend of mobility is faster (62~64% water content). At this time, the fluidized scrapped soil presents a liquid state, see Fig. 4.3d. It can be seen that there is water excretion on the surface of the sample, which indicates that the water content is too high, which is not conducive to later construction use.

The fluidized scrapped soil's water is divided into two parts: fill water and surface layer water. The filling water exists in the pores between the particles, while the surface layer of water exists on the surface of the particles and will form a water film layer on the surface. The flow performance of the fluidized scrapped soil mainly depends on the thickness of the water film formed by the surface layer of water. When steel slag is used instead of cement, due to its small particle size, the filling water in the original void will be replaced to form free water (Nan et al. 2021), which increases the thickness of the water film layer, so the use of steel slag instead of cement will improve its fluidity. Secondly, the particle size distribution of steel slag after grinding is more reasonable. The appearance is spherical, and they are wrapped on the surface of relatively rough cement and scrapped soil to play a certain lubricating role, reducing the internal friction resistance and slightly improving the flow performance.

4.3.2 *Temporality of Fluidized Scrapped Soil at Different Times*

Figure 4.4(a–d) show the flow curves of fluidized scrapped soil with different water content under the four steel slag substitution rates of 50.0%, 58.3%, 66.7%, and 75.0%, respectively. As can be seen from the Figs., with the increase of time, the flow degree under the four water content is decreasing and conforms to the linear

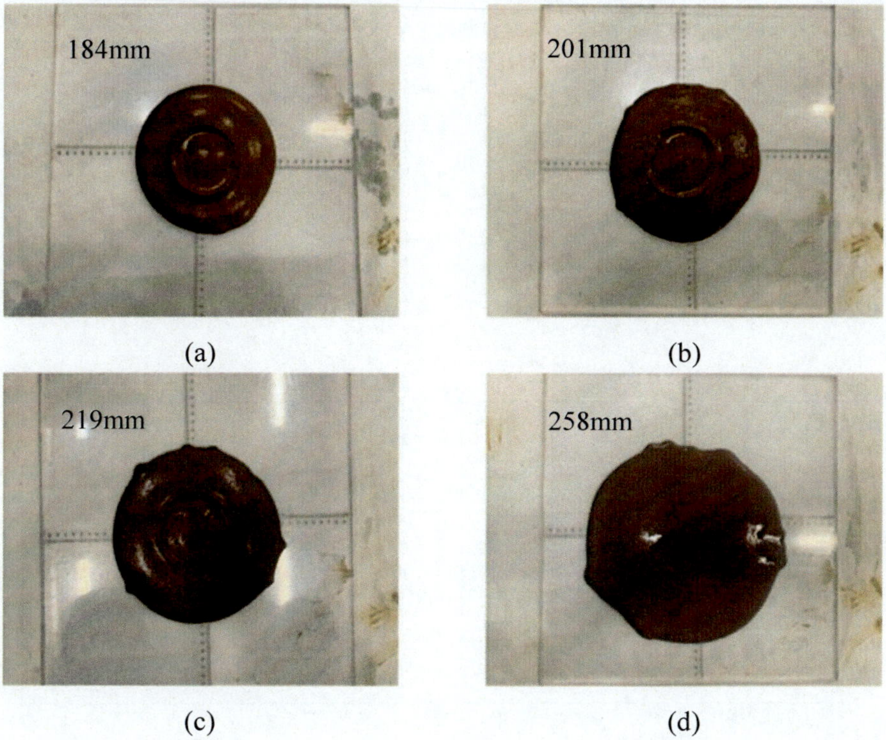

Fig. 4.3 Fluidized scrapped soil with different fluidity

decrease law. According to the fitting curve, the decrease in mobility under the three water content of 58%, 60%, and 62% is the same, which indicates that in the range of water content of 58~62%, the influence of water content on the flow degree is small. If the water content of the mobile soil is continued to be increased on this basis, the excess water in the mixture will be too much, resulting in the mixture becoming more and more dilute, and the amount of cement is not enough to consume these water, and the integrity of the mixture will decrease.

The main reason why the flow properties of fluidized scrapped soil decrease over time is the hydration reaction caused by cement. The reaction of Ca_3O_5Si in the cement used in the study with water produces hydrated calcium silicate, which mainly presents gelatinous particles, which are gradually connected to form a skeleton with a certain strength as these gelatinous particles increase, resulting in a decrease in flow. In addition, the active particles in the scrapped soil will also be cemented with these cementitious particles to reduce the flow degree of the fluidized scrapped soil.

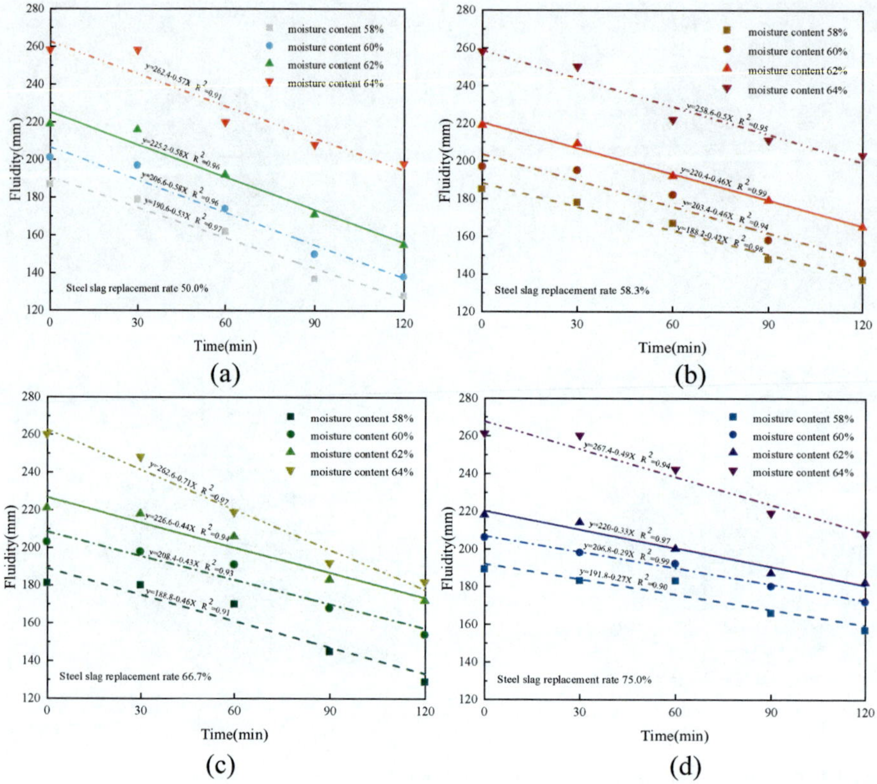

Fig. 4.4 Fluidity curve of fluidized scrapped soil with different water content under four steel slag replacement rates

4.3.3 *Filling Performance of Fluidized Scrapped Soil Under Different Water Content*

According to the results of the above fluidity test and the classification table of the fluidized soil, compared with the backfill with water content in the range of 60~64%, that is, a mobility degree in the range of 200 ~ 260 mm. When the water content is less than 60%, that is, the fluidity is less than 200 mm. Although the backfilling rate of the fluidized scrapped soil also meets the standard, its backfilling effect is slightly worse, which also shows that the filling performance of the fluidized scrapped soil is improved while improving the water content and improving fluidity.

The backfilling situation in the test is mainly divided into Figs. 4.5 and 4.6. Figure 4.5 is the backfilling of fluidized scrapped soil with a water content of 58%, and Fig. 4.6 is the backfilling of fluidized scrapped soil with a water content of 60~64%. From the two Figs, if the backfill area includes both conventional space and special-shaped section structure when the water content is 58%, the left-side

Fig. 4.5 Backfill situation at 58% water content

conventional space backfill is very good, but there are still some areas at the right-hand special-shaped cross-section structure that are not filled. When the water content is 60~64%, the whole backfill area is backfilled well. Although the filling rate of both meets the basic requirements of greater than 80%. But when the water content is 58%, that is, the flow degree is less than 200 mm, the backfilling ability of the fluidized scrapped soil for this special-shaped structure surface is still lacking. Therefore, the fluidized waste slag in the range of 60~64%, that is, the flow degree in the range of 200~260 mm, is more suitable for those foundation pit sidewall projects with special-shaped section structures.

4.4 Conclusion

In this paper, the construction and workability of fluidized scrapped soil are studied. Mainly from the aspects of flowability, timeliness and filling performance, the influence of steel slag substitution rate and water content on its construction and workability was studied. There are mainly the following conclusions:

(1) The use of steel slag instead of cement to prepare fluidized scrapped soil can slow down the rate of decrease of fluidity over time. The degree of mobility shows a gradual decrease with the increase of time.

(2) When the water content and steel slag replacement rate of fluidized scrapped soil meet the following conditions: ① the water content is 60% and the steel slag

Fig. 4.6 Backfill situation at 60~64% moisture content

replacement rate ≤66.7%; ② the water content is 62%, the steel slag replacement rate is ≤58.3%, its water secretion rate is less than 5%, and the water retention performance is good.

(3) When there is a special-shaped cross-sectional structure in the backfill area of the side wall of the foundation pit, the backfilling effect is better when the water content is greater than 58%.

Acknowledgements The authors would like to thank the National Natural Science Foundation of China (Grant No. 42007263), the Key Research and Development Program (Social Development) project of Zhenjiang (Grant No. SH2022017), the Science and Technology Project of the Ministry of Housing and Urban-Rural Development of China (Grant No. 2019-K-136), and the China Postdoctoral Science Foundation funded project (Grant No. 2020M671297) for supporting the research.

References

Achtemichuk S, Hubbard J, Sluce R, Shehata MH (2009) The utilization of recycled concrete aggregate to produce controlled low-strength materials without using Portland cement. Cement Concr Compos 31(8):564–569

American Concrete Institute (1999) Controlled low-strength materials (CLSM). ACI 229

Behiry EM (2013) Evaluation of steel slag and crushed limestone mixtures as subbase material in flexible pavement. Ain Shams Eng J 4(1):43–53

Cheng F, Li H, Li B, Zhang L (2009) Experimental study on the use of low alkalinity converter steel slag as mineral admixture. J Xi'an Univ Architect Technol (Natural Science Edition) 41(6):871–875

Gemperline CS, Durham S (2019) Beneficial use of recycled materials in controlled low strength materials. ICPTT 2012: Better Pipeline Infrastructure for a Better Life 1305–1316

Horiguchi T, Fujita R, Shimura K (2011) Applicability of controlled low-strength materials with incinerated sewage sludge ash and crushed-stone powder. J Mater Civ Eng 23(6):767–771

Huang Y, Liu Z, Liu L, Lan T, Fu Y (2021) Steel slag-cement stable crushed rock performance and environmental impact test. J Chang'an Univ (Natural Science Edition) 41(5):43–53

Lachemi M, Saharan M, Hossain KMA, Lotfy A, Shehata M (2010) Properties of controlled low-strength materials incorporating cement kiln dust and slag. Cement Concr Compos 32(8):623–629

Li P, Ren Q, Lu Q, Chen R (2022) Review of low-carbon cement feedstock/fuel alternative technology for dual carbon. Clean Coal Technol 28(8):35–42

Ling TC, Kaliyavaradhan SK, Chi SP (2018) Global perspective on the application of controlled low strength material (CLSM) for trench backfilling-An overview. Constr Build Mater 158:535–548

Lu Y, Wang X, Qin D, Liu Y, Wang H (2018) Efficient dephosphorization and phosphorus fixation of low alkalinity steel slag in converter steelmaking. Iron Steel 53(6):31–38

Luo L, Guo Y, Li Y, Zhang Z, Zhu T (2022) Research on technical path and prediction of low-carbon development of cement industry under carbon neutrality. Res Environ Sci 35(6):1527–1537

Mason B (1944) The Constitution of some open-heart slags. J Iron Steel Institute 1(1):69–80

Menina A, Soliman A, Ahmed A, Naggar H (2016) Green controlled low-Strength material. Vancouver

Mo J, Zhang W, Chen J, Zeng X, Yang T, Zan J (2019) Comparison of two different alkalinity steel slags and HAP-supported cadmium adsorbed. Chinese J Environ Eng 13(8):1800–1808

Nan X, Yang L, Tang W, Han B, Wang C (2021) Gray analysis of the effect of steel slag particle size distribution on the hydration characteristics of steel slag cement. J Lanzhou Univ Technol 47(5):138–143

Nataraja MC, Nalanda Y (2008) Performance of industrial by-products in controlled low-strength materials (CLSM). Waste Manage 28(7):1168–1181

Sheen YN, Sun TH, Chung WH (2008) Compressive strength of controlled low-strength materials containing stainless steel slag. Corros. Prot. Eng. 22(3):217–229

Thunder (2017) Performance optimization and pilot test study of low-intensity mobile backfill materials for construction waste. Beijing University of Technology

Wang Z (2015) Research on muck improvement technology of underwater section tunnel of Nanchang Metro. Nanchang Institute of Technology

Wang HY, Chen KW (2016) A study of the engineering properties of CLSM with a new type of slag. Constr Build Mater 102(1):422–427

Wu JY, Lee MZ (2011) Beneficial reuse of construction surplus clay in CLSM. Int J Pavement Res. Technol. 4(5):293–300

Yu Y (2019) Experimental analysis of bentonite slurry for shield muck improvement. Shenyang University of Technology

Zhao L, Zhang F (2020) Comprehensive utilization and development prospect of steel slag resources. Mater Rev 34(S2):1319–1333

Chapter 5
Assessment of Green Infrastructures Performance for Water Quality Management

Qian Yu and Na Li

Abstract With the dual pressure of climate change and rapid urbanization, urban pluvial flooding and water quality degradation issues are becoming increasingly serious. Green infrastructure (GI), a sort of resilient practice, is able to effectively improve the water quality of stormwater runoff. When planning and designing GIs that are adapted to both local conditions and future climate change, modeling the effects of GIs on controlling stormwater runoff pollutants under various rainfall characteristics is crucial. In this paper, we set three rainfall scenarios with varying rainfall return periods and then assess the GIs performances on the improvement of water quality in the Jinan Daminghu Sponge City Construction pilot area. The results indicate that GIs have a positive effect on the control of rainfall-runoff pollutants. Improvement effects decrease as rainfall return periods increase. The suspended solid (*SS*) load reduction rates are 68.552%, 67.942%, and 67.314% under the 24-h design storm with a return period of 5, 10 and 20 years. The main reason is the lower rates of reduction of runoff coefficient. The corresponding values are 10.148%, 8.406%, and 6.611% for three rainfall return periods. In addition, we vary the *SS* removal efficiency of each GI and the results show that the control effects on *SS* loads are sensitive to the *SS* removal efficiency.

Keywords Green infrastructure · Water quality · Suspended solid · Performance evaluation

---

Q. Yu (✉) · N. Li
China Institute of Water Resources and Hydropower Research, Beijing 100038, China
e-mail: yuqian@iwhr.com

Research Center on Flood & Drought Disaster Reduction of the Ministry of Water Resources, Beijing 100038, China

H. Xu (ed.), *Proceedings of the 5th International Symposium on Water Resource and Environmental Management*, Environmental Science and Engineering,
https://doi.org/10.1007/978-3-031-31289-2_5

5.1 Introduction

Urbanization has nearly completely taken over China in the previous few decades, causing changes to urban densities, land use, and population growth. Traditional development practices lead to higher rates of imperviousness, which raise surface runoff volumes and lower infiltration rates (Ahiablame et al. 2012). Rapid urbanization has negative effects, including an increase in the risk of urban flooding, deteriorating water quality, environmental harm, and more (Yu et al. 2020). According to the most recent IPCC study (IPCC 2021), Asia is also expected to experience more frequent and intense rainfall (Yu et al. 2020). Compared to the only benefit of grey infrastructure, green infrastructure (GI), a sort of resilient practice, is an effective measure to reduce the negative impacts caused by urbanization and adapt to climate change by storing, detaining, infiltrating, and purifying rainfall runoffs at the source (Dietz 2007).

To simultaneously alleviate the urban flood problems and solve water environmental and ecological water problems, Sponge City Construction (SCC) has been promoted in China since 2013 (Li et al. 2017). The core of SCC is low-impact development (LID) or GI. Compared with the number of studies on GI performances on water quantity, studies on their water quality performance under different rainfall return periods are not sufficient. In addition, there is little research on sensitivity analysis for GI performance on water quality.

Given the above, we first examine the effects of GIs on water quality under various periods of precipitation. We then conduct a sensitivity analysis of the effectiveness of suspended solid (*SS*) removal.

5.2 Methods and Materials

5.2.1 Study Area

With an average yearly rainfall of 672.8 mm, Jinan, one of China's first 16 national pilot SCC cities, is situated in the east of the country. With a total area of around 39 km^2, the study area is situated in Jinan's major urban regions. The south and east of the study region have higher topography than the north and west, which have lower terrain. Mountainous landscapes are in the eastern and southern sections, while the piedmont slope lies in the middle. Because of the geography of the research area, severe flood disasters could occur in the summer during periods of heavy precipitation.

According to the 2015–2017 Sponge City Construction Pilot Project Implementation Plan in Jinan, Shandong Province (hereinafter referred to as the implementation plan). Five kinds of GIs are designed within the study area: Green roofs (0.26 km^2), sunken greenbelt (0.31 km^2), retained greenbelt (1.32 km^2), intensified infiltration greenbelt (0.46 km^2), and permeable pavement (0.18 km^2) (see Fig. 5.1).

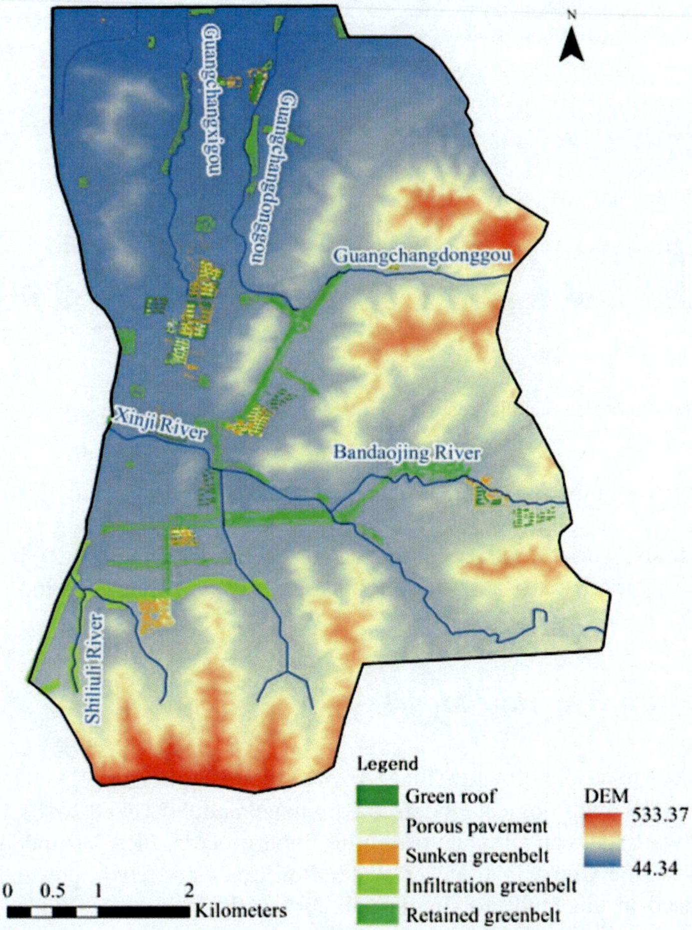

Fig. 5.1 Distributions of GIs in the study area

5.2.2 Data

The following basic geographical data are used for scenario modeling, i.e., DEM data, DLG data 1:2000, and soil data 1:1,000,000 (1995). DLG data includes spatial distributions of roads, residential areas, and land uses. The above data adopts the CGCS2000 coordinate system, Gauss Kruger's projection, and the coordinate unit is one meter. The 1985 national elevation benchmark is adopted.

In addition, the hydrological and meteorological data are also collected, i.e., the river systems, the main river cross sections, the monitoring precipitation data at Guishan, Xinglong, and other precipitation or hydrological stations, and the design storm data as well, including 5-year return period, 10-year return period and 20-year return period. Moreover, data on drainage systems are also used. Currently, the

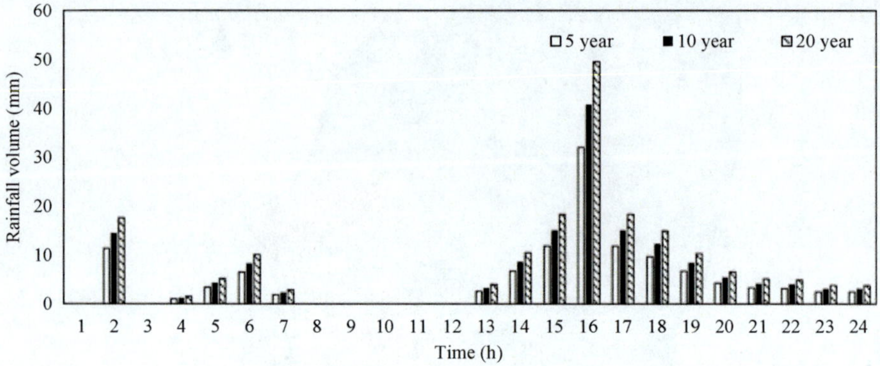

Fig. 5.2 The 24 h design rainfall process under three different rainfall frequency scenarios

drainage system in the study area is designed to have a return period of two to three years.

In this study, we set three precipitation return periods (see Fig. 5.2), including a 5-year return period, a 10-year return period, and a 20-year return period.

5.2.3 Hydro-Hydraulic Model

Flood Risk Analysis Software (FRAS), mainly including the 1D–2D coupling hydraulic model, hydrological model, and drainage model (Li et al. 2018), is an integrated software that can simulate the entire flood process. FRAS is independently developed by the China Institute of Water Resources and Hydropower Research and it is used in this study to simulate the urban flood processes under different rainfall return periods. The SCS-CN model is selected in this research to simulate rainfall-runoff generation. Furthermore, the equivalent piping system model is used to simulate underground drainage.

An irregular grid with a total number of 26,659 is divided with an average grid area of 1500 m^2. The rivers and roads are set as special passages. Roughness coefficients are defined in relation to land use types. Five types of GIs are modified by parameterizing, i.e., elevations, roughness, and values of CN (Yu et al. 2021). Retained greenbelt is also considered as an intensified infiltration greenbelt. Details about calibration and model validation are available in Li et al. (2018), which are not elaborated on in this paper.

5.2.4 Water Quality Analysis

Suspended solid (*SS*) is one of the main pollutants in urban areas. Some types of GIs can effectively remove *SS* from stormwater runoffs. The *SS* is therefore chosen to represent surface water quality in this study. The empirical method developed by Schueler (1987) is used to estimate pollution loads from stormwater runoff in urban areas.

$$L = C_F \cdot \varphi \cdot A \cdot P \cdot C \tag{5.1}$$

where L is the *SS* loads in the runoffs, kg; C_F is the factor to correct rainfall events that do not produce surface runoffs, i.e., the proportion of rainfall events that produce runoffs to the total rainfall events within one year, if 90% rainfall events within one year produce runoffs and then $C_F = 0.9$, and if estimate loads of one rainfall event and then $C_F = 1.0$; φ is the average runoff coefficient of the study area, which is calculated by FRAS in this study; A is the catchment area, km^2; P is the rainfall volumes, mm; C is the weighted average concentration of *SS*, mg/L.

Compared with available data on *SS* values in the United States, most Chinese cities still do not have concentration values for storm runoff pollutants. During precipitation processes, concentrations of pollutants in runoff vary significantly with precipitation characteristics. In addition, pollutant concentrations vary widely depending on land use in the study area. Hence, the event mean concentration (*EMC*) values are used in this study, which is usually measured by laboratory analysis of flow- and time-weighted composite samples. *EMCs* are the most common parameters used to estimate non-point pollutant loads, which can be calculated according to Eq. (5.2).

$$EMC = \frac{\sum_{j=1}^{n} C_j V_j}{\sum_{j=1}^{n} V_j} \tag{5.2}$$

where C_j is the measured pollutant concentration at j^{th} time period, g/m^3; V_j is the runoff volume at the jth time period, m^3; n is the number of time segments. Then Eq. (5.1) can be reformulated into Eq. (5.3).

$$L = C_F \cdot \varphi \cdot A \cdot P \cdot EMC \tag{5.3}$$

The removal efficiency of *SS* loads is an important index to evaluate the GI performance on runoff water quality. However, it is difficult to determine both the SS loads within the study area and the *SS* removal effectiveness of each GI. First, the concentration of *SS* differs significantly between cities and land uses. Moreover, the values vary with different measured times, even within the same city. Besides, the removal efficiency varies with different kinds of GIs and also varies with different designs and materials of the same GIs.

Given the lack of long-term measured data before and after the implementation of GIs in the study area, we choose the monitored data from a similar area (Beijing).

Table 5.1 *EMCs* of *SS* of different land use (mg/L)

	Road	Green space	Roof	Others
EMCs	297.37	320.68	28.64	183.68

Table 5.2 The removal efficiency of different GIs on *SS* (%)

Lawn	Rain garden	Vegetated filter strips	Drainage pipeline	Swales	Detention ponds
0	75	65	0	65	60

The values are listed in Table 5.1 (Yang et al. 2015). As *EMC* values vary by land use, the *EMC* of *SS* is calculated using the weighted average method. The *EMC* of *SS* is 243.4 mg/L in this study.

Table 5.2 lists the removal efficiency on *SS* of different GIs, which is provided by Green Values developed by USEPA. According to Table 5.2, we set the removal efficiency of *SS* to 65% in this study. In addition, to analyze the effects of control effects of each GI, we vary the removal efficiency to 45.5% (−50%) and 84.5% (+50%).

5.3 Results and Discussion

5.3.1 *Effects of GIs on Runoff Coefficients*

Table 5.3 shows the control effects of GIs on runoff coefficients under different rainfall return periods. The results show that the runoff coefficients after the implementation of GIs are lower than those before the implementation, which means that GIs can control rainfall runoff to some extent. In addition, as the rainfall return periods increase, the corresponding runoff coefficients increase. The runoff coefficients under 5-year, 10-year, and 20-year events are 0.465, 0.512, and 0.540, respectively. The simulation results are in line with the field investigations conducted by Carpenter and Kaluvakolanu (2010). They found that the average runoff coefficient on green roofs is 0.044 under small rainfalls (<12.7 mm), 0.131 under middle rainfalls (12.7 ~ 25.4 mm), and 0.591 under heavy rainfalls (>25.4 mm), based on 21 rainfall events. With the increases in rainfall return periods, the reduction ratios of control effects decrease obviously. The reduction ratios of control effects under 5-year, 10-year, and 20-year events are 10.148%, 8.406%, and 6.611%, respectively. The simulation results are in agreement with those of Yin et al. (2021).

a. **Effects of GIs on surface water quality under different return periods**

Table 5.4 shows the control effects on *SS* loads under different rainfall return periods. The results show that the pollution loads of *SS* after the implementations of GIs are much less than that before the implementations. GIs can effectively reduce the *SS*

Table 5.3 Control effects on runoff coefficient under different rainfall return periods

No	Scenarios	Runoff coefficient		Reduction ratio (%)
		Before	After	
1	5-year	0.517	0.465	10.148
2	10-year	0.559	0.512	8.406
3	20-year	0.578	0.540	6.611

Table 5.4 Control effects on *SS* load under different rainfall return periods

No	Scenarios	*SS* loads		Reduction ratio (%)
		Before (kg)	After (kg)	
1	5-year	629,153.175	197,857.531	68.552
2	10-year	866,897.310	277,909.004	67.942
3	20-year	1,088,380.785	355,751.258	67.314

Table 5.5 Control effects on *SS* load under different removal efficiencies

No	Removal efficiency (%)	*SS* loads		Reduction ratio (%)
		Before (kg)	After (kg)	
1	65	629,153.175	197,857.531	68.552
2	45.5	629,153.175	308,092.441	51.031
3	84.5	629,153.175	87,622.621	86.073

loads and improve rainfall runoff water quality. The simulation results are consistent with previous research. As the return periods increase, the corresponding reduction ratios decline. Declining trends are in line with runoff coefficients.

b. **Effects of GIs on surface water quality under different removal efficiencies**

Table 5.5 shows the control effects on *SS* loads with different removal efficiencies of GIs under a 5-year return period. The results show that removal efficiencies of GIs play an important role in the reduction rates of GIs on *SS* loads. As a result, it is important to design cost-effective measurements based on the characteristics and topography of local precipitation.

5.4 Conclusions

GI performances towards water quality management under different return periods are studied in the Jinan Daminghu Sponge City Construction pilot area. Control effects on runoff coefficients and water quality are reduced with increasing return periods. The effects on water quality are much more obvious than those on runoff coefficients. In addition, the control effects on *SS* loads are sensitive to the removal efficiency of each GI.

Acknowledgements This work was supported by the National Natural Science Foundation of China [No. 51909273] and the Talent Innovation Team for the Strategic Research on Flood and Drought Disaster Prevention of the Ministry of Water Resources [No. WH0145B042021]. The support provided by the IWHR Talented International Experts Program is also acknowledged.

References

Ahiablame LM, Engel BA, Chaubey I (2012) Effectiveness of low impact development practices: literature review and suggestions for future research [J]. Water Air Soil Pollut 223(7):4253–4273

Dietz ME (2007) Low impact development practices: A review of current research and recommendations for future directions [J]. Water Air Soil Pollut 186:351–363

IPCC (2021) Summary for policymakers. In: Climate change 2021: The physical science basis. contribution of working group I to the sixth assessment report of the intergovernmental panel on climate change [Masson-Delmotte, V., P. et al (eds.)]. Cambridge University Press. In Press

Li H, Ding LQ, Ren ML, Li C, Wang H (2017) Sponge city construction in China: a survey of the challenges and opportunities. Water 9(9):594

Li N, Meng YT, Wang J, Yu Q, Zhang NQ (2018) Effect of low impact development measures on inundation reduction-Taking Jinan pilot area as example. J Hydraul Eng 49(12):1489–1502 (in Chinese)

Schueler TR (1987) Controlling urban runoff: A practical manual for planning and designing urban BMP's. Metropolitan Information Center, Metropolitan Washington Council of Governments, Washington, DC

Yang L, Sun CH, Wang YG, Fan Q, Liu GZ (2015) Construction of the dynamic update system of the pollution load of urban non-point sources. Environ Prot Sci 41(2):63–66

Yin DK, Chen ZX, Li Q, Jia HF, Liu ZQ, Shen L, Ahmad S (2021) Influence of rainfall characteristics on runoff control of a sponge reconstructed community in a rainy city. J Tsinghua Univ (Sci Technol) 61(01):50–56 (in Chinese)

Yu Q, Li N, Wang S, Meng YT (2020) Study on comprehensive benefit assessment systems for low impact development practices. J Hydroelectr Eng 39(12):94–103 (in Chinese)

Yu Q, Du XH, Li N, Meng YT, Wang J (2021) Modelling the effects of green infrastructures on water quantity under different rainfall characteristics. 2nd Int Symp Water, Ecol Environ

Chapter 6
Removal Rate of Pollutants in Water Body by MBBR Based on Folded Cylindrical Biological Carrier

Jian Zhu, Hui Tan, and Rui Wang

Abstract MBBR process is the effective mean to improve the water quality in rivers and lakes, and the folded cylindrical biological carrier has been proved to have a high specific surface area and good hydrodynamic properties. In this study, the removal rate of the pollutants in water body by MBBR based on the folded cylindrical biological carrier were experimentally investigated in the self-developed MBBR equipment of Harbor and Coastal Engineering Laboratory of Shanghai Jiao Tong University. A series of experimental runs were conducted with different pollutants, and the experiment setup was introduced in detail. And then, the calculation model for the removal rate of pollutants was derived. Based on the experimental date and field measured data, the treatment effect of MBBR based on folded cylindrical biological carrier in polluted water body was monitored, and the undetermined coefficients of the calculation model were determined. The research results indicated that the pollutant indexes (COD, NH_3–N and TP) in polluting water bodies decreased steadily, but the removal rate was different after adding the folded cylindrical biological carriers into the biological reaction tank of MBBR; the calculation method proposed in this paper could accurately obtain the treatment rate of pollution indexes in polluting water body. The research results provided a reference for the future engineering practice of MBBR based on the folded cylindrical biological carrier.

Keywords Removal rate of pollutants · MBBR · Folded cylindrical biological carrier · Treatment effect · Calculation model

J. Zhu · H. Tan
Guangxi Taineng Engineering Consulting Co. Ltd, Guangxi 53000, China

R. Wang (✉)
School of Naval Architecture, Ocean and Civil Engineering, Shanghai Jiao Tong University, Shanghai 200240, China
e-mail: w530707596@sjtu.edu.cn

H. Xu (ed.), *Proceedings of the 5th International Symposium on Water Resource and Environmental Management*, Environmental Science and Engineering,
https://doi.org/10.1007/978-3-031-31289-2_6

6.1 Introduction

With the development of social economy and the improvement of people's living quality, the requirements for ecological environment are increasing, and the comparable requirements for the indicators of sewage discharge are becoming more and more strict (Mczaowski 2019; Shuvalov 2020). In China, the standard of First-Class A has been proposed by the ministry of environmental protection of the people's republic of China in 2006, but it can't no longer meet the actual needs of some places after more than a decade of development (Peng et al. 2021), such as Beijing, Tianjin, Taihu and so on. In order to cope with the increasingly stringent requirements for sewage discharge indicators, it is urgent to conduct more in-depth research on the process of sewage treatment.

MBBR process is an effective means for sewage treatment (Leyva-Diaz et al. 2016a), and a lot of excellent results have been achieved, such as excellent results about the performance (Leyva-Diaz and Martin-Pascual 2015), the membrane fouling (Zhang et al. 2020), the hydraulic residence time (Rodriguez-Sanchez et al. 2018), the application of MBBR process in different water body (Biswas et al. 2014; Schopf et al. 2018; Tsitouras et al. 2022) and so on. The treatment efficiency of MBBR process has been concerned by many scholars, but most of their study mainly focused on the treatment technology process (Leyva-Diaz et al. 2016b), the improvement of the optimization of microbial strains (Hu et al. 2013), and the modify of biological carriers (Deng et al. 2016; Mahdavi and Norouzian 2018), but there is no clear method to predict its treatment efficiency.

Biological carrier is the important part of MBBR process, and many scholars have studied the manufacturing technology (Wang 2018), the constituent materials (Tang et al. 2018), the structure (Ma et al. 2014; Iordache et al. 2020), and the hydrodynamic properties (Ogunsona et al. 2017) of biological carriers accordingly. The folded cylindrical HDPE biological carrier is a new type of biological carrier, which is proposed by Wang et al. (2022) and they have conducted a detailed analysis of the hydrodynamic properties and the structure of this biological carrier, and completed the investigation of its treatment effect in the laboratory, but unfortunately didn't give the practical effect in MBBR process. In addition, the removal efficiency is also not analyzed in detail (Wang et al. 2022).

In this study, the removal rate of pollutants in water bodies by MBBR process based on the folded cylindrical biological carriers was investigated by conducting the laboratory experiments and the field tests, and the application effect of the folded biological carriers in MBBR process was specifically analyzed accord to the experimental results and field test results. On this basis, a calculation model to predict the removal rate of pollutants in water bodies by MBBR based on folded cylindrical biological carriers was proposed for practical reference.

6.2 Materials and Methods

6.2.1 Test Materials

The folded cylindrical HDPE (High Density Polyethylene) biological carriers with 36 wells were selected in this study. It has been found that the biological carriers used in MBBR are mostly composed of HDPE (High Density Polyethylene), and have a central symmetric structure, which have the characteristics of light weight, heat resistance and strong reoxygenation ability, etc. The folded cylindrical biological carriers with 36 wells prepared by HDPE are a new type of biological carriers, which are developed by Shanghai Jiao Tong University, and produced by Wuxi Haituo Environmental Protection Equipment Co., Ltd., as shown in Fig. 6.1. The outer edge and inner support structure of this biological carrier are both the folded structure. Wang et al. (2022) have proved that this type of biological carrier has a high specific surface area and good hydrodynamic properties. Compared with the biological carriers with the same pore number, the specific surface area of the folded cylindrical biological carrier is increased by about 5.1%, and the average gas holdup in the reactor tank can be increased by about 20%–30% by adding the folded cylindrical HDPE biological carriers in to the MBBR system (Wang et al. 2022).

6.2.2 Experiment Setup

The experiment setup of the self-developed MBBR equipment is shown in Fig. 6.2. The reaction tank was made of the transparent acrylic with the inlets of airflow and water flow. The height and the diameter of the reaction tank are both 80 cm. One aeration disc with the diameter of 21.5 cm was installed at the bottom of the reaction

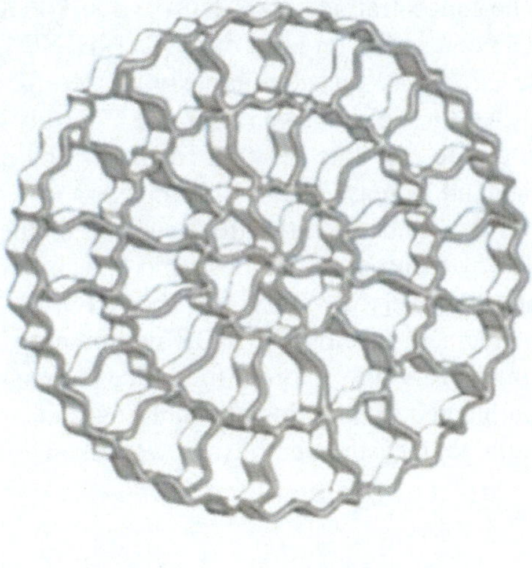

Fig. 6.1 Folded cylindrical biological carrier with 36 wells prepared by HDPE

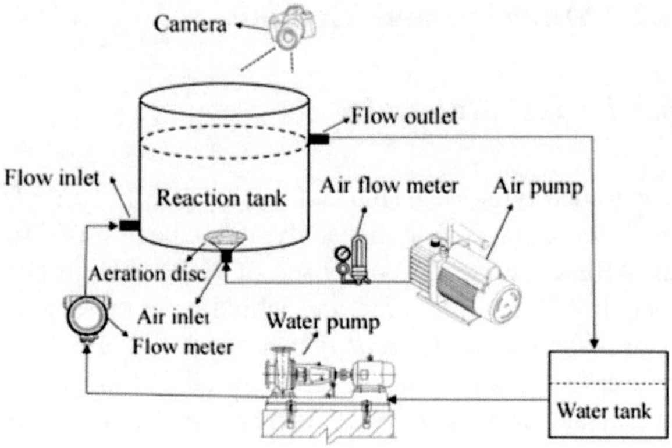

Fig. 6.2 Experiment setup of self-developed MBBR equipment

tank, and connected with the inlet of airflow to release the airflow produced by the air pump. One gas flow meter was used to monitor the speed of the airflow, which was installed between the air pump and the aeration disc. One submersible pump was used to control the flow rate which was measured by the flow meter installed between the submersible pump and the water inlet. The movement of the biological carriers in the reaction tank was record by the high frame rate camera to ensure its full fluidization. The recycled water was stored by the water tank.

6.2.3 *Biofilm Culture*

The method of artificial inoculation was adopted to perform the biofilm culture. Before biofilm culture, the activated sludge was added into the reaction tank in advance, and the concentration of COD in the culture solution was controlled at 300 mg/L. After that, NH_4NO_3 and KH_2PO_4 were added into the culture solution. The concentration of NH_4NO_3 and KH_2PO_4 were 15 mg/L and 3 mg/L respectively. The concentration ratio of COD, NH_4NO_3 and KH_2PO_4 was 100:5:1. Thirdly, the reaction tank was continuously aerated for 24 h, and then the upper supernatant of the culture solution was removed. Fourthly, the same amount of culture solution was added to continue aeration for 24 h. The inflow rate was set at 6 L/min at the 1st day and increased by 2/3 L/min every day. The inflow rate reached 10 L/min at the 6th day. Subsequently, the flow rate of the inflow was kept constant for 10 days. The concentrations of COD and NH_3–N of the inflow and outflow were measured at 4 PM every day during the period of the biofilm culture and the stable operation of the self-developed MBBR equipment. Each measurement was carried out 3–4 times every day, in order to reduce the unpredictable uncertainties produced by the pollutant concentration, the nutrients and the microorganisms, and the average of the multiple measurement results was taken as the measurement data.

6.2.4 Field Test

The folded cylindrical HDPE biological carriers and the MBBR system produced by Wuxi Haituo Environmental Protection Equipment Co., Ltd. were used in the treatment process of the aquaculture tail water discharged from the breeding pond of the hydropower station. The continuous change of the pollutant indicators, such as COD, BOD, NH_3–N, TP and dissolved oxygen in the inflow and outflow of the MBBR system were monitored for 21 consecutive days. The aquaculture tail water in this place was inferior to fifth class standard before treatment and the content of each contaminant is shown in Table 6.1. The MBBR equipment is shown in Fig. 6.3.

The treatment technology process of the MBBR system in the in the treatment process of the aquaculture tail water is as follows. Firstly, the aquaculture tail water was collected in the catchment tank, and then lifted into the regulating tank by water pump. The regulating tank borne the fluctuations in the amount of water caused by the irregular drainage, and made the aquaculture tail water flow in the reaction tank homogeneous. Subsequently, the aquaculture tail water was discharged into the reaction tank and the delayed aeration was carried out to convert the ammonia nitrogen in aquaculture tail water into the nitrate nitrogen. By returning the sewage to the anaerobic pool, the purpose of the complete denitrification and denitrification was achieved. The MBBR system was equipped with an inclined pipe sedimentation tank, which was divided into a series of shallow sedimentation layers by using inclined parallel pipes, and the treated and settled sedimentation sludge moved and separated

Table 6.1 Water index of inlet and outlet

Index	COD (mg/L)	BOD (mg/L)	NH_3–N (mg/L)	TP (mg/L)	Dissolved oxygen (mg/L)
Inflow	12–35	1.5–4.5	0.2–0.35	0.1–0.28	5.5–8.5

Fig. 6.3 MBBR produced by Wuxi Haituo Environmental Protection Equipment Co., Ltd

in each shallow sedimentation layer. Finally, the aquaculture tail water after treatment by the MBBR system was collected into the reservoir to be reused.

6.3 Results and Discussion

6.3.1 Experimental Result

During the experiments, the hydraulic residence time was set as 0.015 day, and the aeration rate was set as 10 L/min. The filling rate of the reaction tank was 0.5. The specific surface area of the folded cylindrical HDPE biological carrier with 36 wells was 662.5 m^2/m^3, and the material density of the biological carrier was 0.927 kg/m^3. The water temperature, the concentration of COD, NH_3–N and TP in the inflow and out flow of the self-developed MBBR equipment was measured, and the experimental results are shown in Table 6.2.

6.3.2 Field Test Result

The concentrations of COD in the inflow and outflow of MBBR system are shown in Fig. 6.4. It could be seen that the removal effect of COD was significant on the 10th day, because the biological carriers were during the period of the biofilm culture from the 1st day to the 9th day, and the biofilm culture finished on the 10th day. The MBBR system began to work normally from the 10th day, and the subsequent value of COD concentration was basically stable at 15–20 mg/L, which had reached the III Class Standard of surface water.

The concentrations of BOD in the inflow and outflow of MBBR system are shown in Fig. 6.5. It could be seen that the removal effect of BOD began to appear on the 5th day, during which the biofilm culture process initially completed; from 6th day to 17th day, the concentrations of BOD were basically stable at about 1 mg/L. From the 18th day, due to the continuous increase of the BOD concentration, the microbial community appeared the short-term inadaptability, but after a few days of the self-adjustment of the microbial community, the concentrations of BOD in the outflow of the MBBR system tended to be stable and could stably reach the II Class Standard of surface water.

The concentrations of NH_3–N in the inflow and outflow of MBBR system are shown in Fig. 6.6. It could be seen that the concentration of NH_3–N fluctuated greatly in the early stage, which had a certain impact on the growth and hanging of the nitrifying bacteria and the denitrifying bacteria; the removal effect of MBBR system on NH_3–N was not reflected from the 1st day to the 9th day, which may be due to the high concentration of NH_3–N inhibiting the microbial activity of the autotrophic nitrifying bacteria; from the 10th day, the concentrations of NH_3–N

Table 6.2 Experimental results

Time (d)	Water temperature (°C)	COD (mg/L)		NH_3–N (mg/L)		TP (mg/L)	
		Inflow	Outflow	Inflow	Outflow	Inflow	Outflow
1	16	63	58	6.21	6.11	0.22	0.11
2	21	61	55	7.11	6.99	0.19	0.09
3	18	69	53	7.02	7.98	0.17	0.09
4	17	65	50	6.98	6.78	0.21	0.1
5	18	66	45	6.87	6.51	0.23	0.11
6	16	63	37	7.23	6.21	0.21	0.09
7	18	67	35	6.99	6.01	0.18	0.12
8	15	75	23	8.01	6.26	0.24	0.12
9	13	69	24	7.12	5.97	0.23	0.09
10	16	63	18	6.67	4.31	0.18	0.08
11	15	78	21	6.92	3.01	0.25	0.13
12	17	71	17	7.01	1.03	0.24	0.11
13	15	68	15	6.88	0.81	0.23	0.09
14	16	65	18	6.91	0.71	0.21	0.11
15	14	69	17	7.12	0.53	0.23	0.15
16	19	82	25	8.13	0.52	0.27	0.09
17	26	79	22	7.13	0.41	0.19	0.11
18	20	76	20	7.01	0.39	0.18	0.09
19	23	89	21	8.11	0.38	0.19	0.12
20	19	78	23	7.98	0.39	0.17	0.09
21	17	71	23	7.17	3.95	0.35	0.09
22	18	70	19	7.21	2.18	0.32	0.17
23	16	69	19	7.03	1.77	0.17	0.15
24	18	74	15	6.91	0.87	0.27	0.18
25	15	72	17	7.03	0.79	0.28	0.09
26	13	71	16	8.11	0.61	0.39	0.12
27	16	69	15	6.91	0.53	0.27	0.09
28	15	78	17	7.12	0.61	0.32	0.11

tended to be stable, and the concentration of NH_3–N in the outflow of MBBR system were maintained at about 0.1–0.2 mg/L, which reached the II Class Standard of surface water.

The concentrations of TP in the inflow and outflow of MBBR system are shown in Fig. 6.7. It could be seen that the condition for the film hanging of the polyphosphorus bacteria was similar to that of the nitrifying bacteria and the denitrifying bacteria; the MBBR system began to work normally on the 17th day, and the concentrations of TP in the outflow of the MBBR system were basically between 0.15–0.20 mg/L,

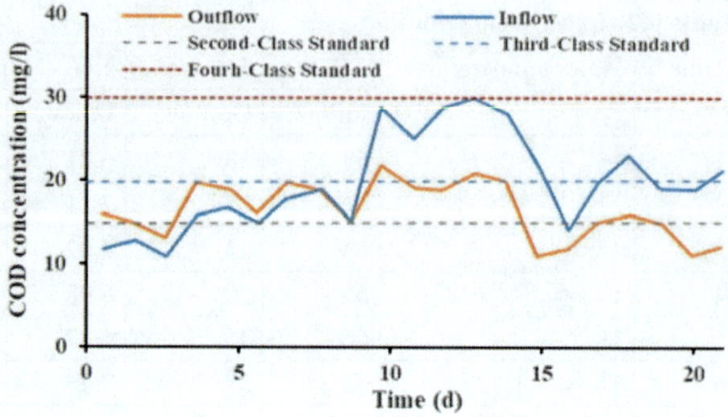

Fig. 6.4 Concentrations of COD in inflow and outflow of MBBR system

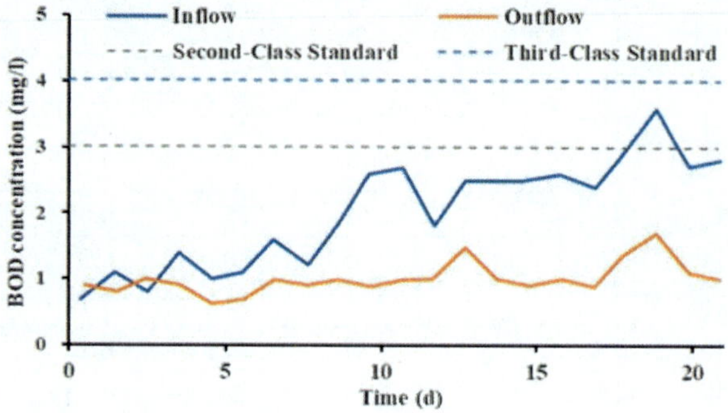

Fig. 6.5 Concentrations of BOD in inflow and outflow of MBBR system

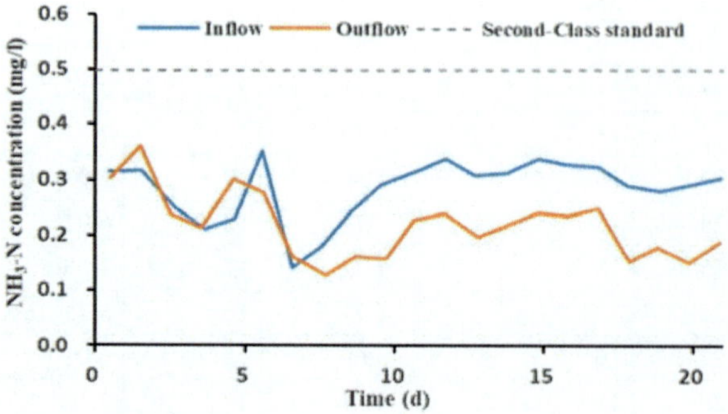

Fig. 6.6 Concentrations of NH_3–N in inflow and outflow of MBBR system

which reached the III Class Standard of surface water. It is worth noting that the concentration of TP in the aquaculture tailwater is already very low, and the MBBR system can still play a removal effect.

Dissolved oxygen is a major limiting factor affecting the normal operation of MBBR system. The concentrations of the dissolved oxygen in the inflow and outflow of MBBR system are shown in Fig. 6.8. It could be seen that the dissolved oxygen in the inflow was higher than 6 mg/L from the 1st day to the 10th day, which met the II Class Standard of surface water, but it is hard to form a hypoxic zone inside

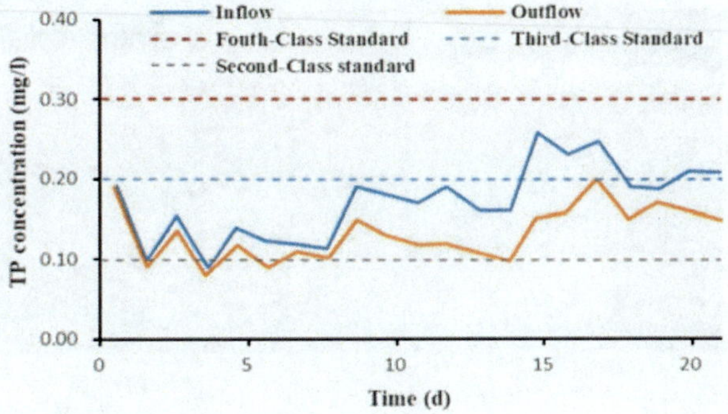

Fig.6.7 Concentrations of TP in inflow and outflow of MBBR system

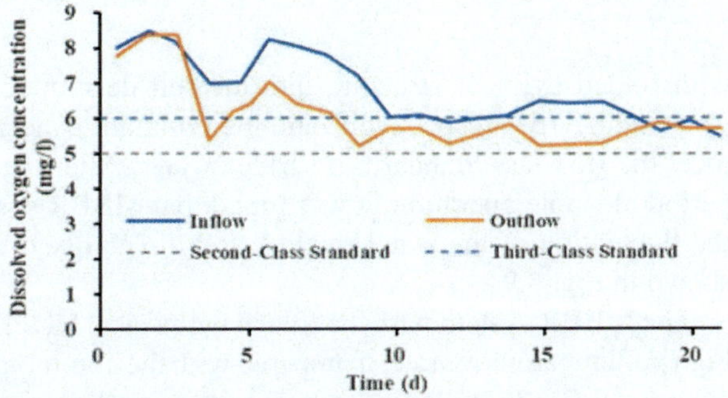

Fig. 6.8 Concentrations of dissolved oxygen in inflow and outflow of MBBR system

the biological carrier due to the high concentration of the dissolved oxygen in the inflow, resulting in the unnormal operation of the MBBR system; before the 10th day, the concentration of the dissolved oxygen in the reaction tank fluctuated greatly and was slightly lower than the concentration of the dissolved oxygen in the inflow, which could still meet the III Class Standard of surface water, because the process of microbial growth and film hanging required a part of the oxygen consumed; the concentration of the dissolved oxygen in the outflow began to stabilize on the 10th day; from the 15th to the 18th, due to the increase of the dissolved oxygen content, the aerobic microorganisms such as polyphosphorus bacteria and nitrifying bacteria could further multiply, resulting in the increase of oxygen consumption and the reduction of the dissolved oxygen concentration in the outflow. In general, the dissolved oxygen concentration in the outflow could be stable between 5–6 mg/L.

6.3.3 Discussion on Removal Effect

MBBR system with the folded cylindrical HDPE biological carrier possesses good pollutant remove effect. Through laboratory experiments, it was found that the average removal rate of COD and NH_3–N by adding the HDPE biological carriers

(a) Inflow (b) Outflow

Fig. 6.9 Clarity of the inlet and outlet water

with folded cylinder structure into the self-developed MBBR equipment reached 95% and 87.75% respectively, and the pollutant concentration in the outflow could meet the III Class Standard of surface water. After the MBBR system entering the period of stable operation, it was found that MBBR system had strong resistance to the fluctuation of the water quality, and the clarity of the inlet and outlet water is shown in Fig. 6.9.

The MBBR system with the folded cylindrical HDPE biological carrier possesses high pollutant remove rate. Compared with the commonly used biological, the folded cylindrical HDPE biological carrier with 36 well used in this study has clear advantages in terms of pollutant treatment efficiency, and basically completes biofilm culture on 17th day. The removal effect of COD by the YL-II. aerobic active biological carrier is significantly enhanced after the 30th day (Huang et al. 2022); The plaque of the K3 biological carrier begins to form at 14th day, and the relatively complete biofilm doesn't form until the 26th day (Sun et al. 2021); a yellow–brown biofilm can" be visual on the surface of the polypropylene hollow sphere type of biocarrier carrier until the 20th days (Wang et al. 2020).

6.4 Calculation Model Derivation

6.4.1 Equilibrium Equation for Pollutant Content Processed by Bio-carriers

In order to analyze the total amount of pollutants in the reaction tank, the following parameters were assumed accordingly. Figure 6.10 shows the diagram of reaction tank filled biological carriers, the fill rate φ of the reaction tank could be derived as follow.

$$\varphi = W/V \tag{6.1}$$

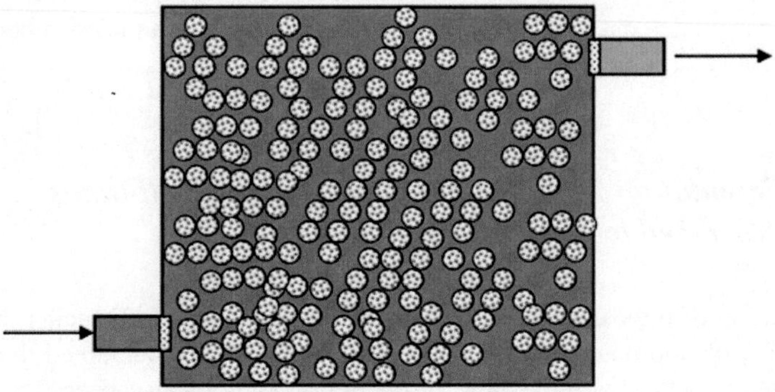

Fig. 6.10 Diagram of reaction tank filled biological carriers

where V was the volume of the reaction tank, m^3; W was the amount of biocarriers filled in the reaction tank, m^3. It was assumed that the water flowing into and out of the reaction tank was the constant flow, and the hydraulic residence of the biological reaction system could be derived as follow.

$$T = \frac{V - W}{Q} = \frac{1 - \varphi}{\varphi Q} W \tag{6.2}$$

where Q was the flow rate, m^3/d.

During a hydraulic residence time, the amount of the pollutants processed by the biological carriers consists of two components: the change of pollutants in the reaction tank and the difference of pollutants content in the inflow and outflow. In addition, the amount of the pollutants processed by the biological carriers is equal to the amount of biological carriers in the reaction tank multiplied by the removal rate of pollutants processed by per unit volume of the biological carriers. Thus, the equilibrium equation could be derived as follow.

$$W\lambda_t T = TQC_{in,t} - IQC_{out,t} + (C_{a,t} - C_{a,t-1})(V - W) \tag{6.3}$$

where λ was the remove rate of the pollutants processed by per unit volume of biological carriers per unit time, kg/m^3/d; $C_{in,t}$ and $C_{out,t}$ were the concentrations of the pollutants flowing into and out of the reaction tank in the time of t respectively, g/m^3; $C_{a,t}$ was the average pollutant concentrations in the reaction tank in the time of t, g/m^3.

According to the above equilibrium equation, the removal rate of the pollutants processed by per unit volume of biological carriers could be derived as follow.

$$\lambda_t = \frac{C_{in,t} - C_{out,t}}{W} Q + \frac{(C_{a,t} - C_{a,t-1})(1 - \varphi)}{\varphi T} \tag{6.4}$$

And the average concentration of pollutants in the reaction tank can be approximated by the equation $C_{a,t} = (C_{in,t} - C_{out,t})/2$.

6.4.2 Calculation Model for Removal Rate of Pollutants Processed by Bio-carriers

The derivation of the calculation model should consider the influencing factors on removal rate of pollutants processed by bio-carriers, which contained the specific surface area and the hydrodynamic properties of biological carrier. The specific surface area and the wells number of biological carriers are corresponding. The hydrodynamic properties of the biological carrier are related to the density of the manufacturing material. For the biological carriers made of HDPE materials, the density is constant, and the removal rate of pollutants is inversely proportional to the hydraulic residence time.

The removal rate of the pollutants processed by per unit volume of biological carriers is undoubtedly also related to the kinematic viscosity of the water body. The kinematic viscosity reflects the scale of the flow resistance of this fluid under the action of gravity, and represents the measure of the internal friction force of the liquid under the action of gravity. The dynamic viscosity can be calculated by using the empirical formula.

$$\nu = \frac{1.775 \times 10^{-6}}{1 + 0.0337 T_o + 0.000221 T_o^2} \tag{6.5}$$

where ν is the dynamic viscosity, m^2/s; T_o is the temperature of water body, °C.

In addition, the temperature of water body is an important parameter to affect the removal rate of the pollutants processed by biological carriers, because the temperature of water body has a great influence on the reproduction rate and reaction rate of microorganisms. Under the same pollution load, the water quality of the out flow increases with the temperature of water body increasing, and the formula of the temperature coefficient is in the form of an exponential function.

For the interaction between incompressible flow and biological carriers, the fluid density ρ, fluid kinematic viscosity ν and hydraulic residence time T were selected as the basic physical quantities. Then the length dimension [L], the mass dimension [M] and the time dimension [T] were the basic dimensions, and the other physical dimensions (water temperature θ, filling rate φ, specific surface area $\mathcal{E}$) are derived dimensions. According to the π theorem and the dimensional analysis, the mathematical expressions for these unit dimensions are developed.

$$\lambda = \beta(1 - \varphi)^{x1} \theta^{x2} (\rho^2 \nu^{1/2} \varepsilon)^{x3} \tag{6.6}$$

where β, x_1, x_2, x_3 were the undetermined coefficients.

It should be noted that the hydraulic residence time in this paper was the nominal hydraulic residence time, which was calculated by the effective volume of reaction and inlet flow. Due to the membrane separation technology of MBBR process, the microorganisms were completely blocked in the reaction tank, and the complete separation of hydraulic retention time and sludge age was achieved. Therefore, the hydraulic residence time didn't appear in the calculation model.

6.4.3 Undetermined Parameters Determination

The coefficients of Eq. (6.6) were determined respectively by using the test results and the collected data with the help of the software of Origin 2021 and Microsoft Excel. For the removal rate of NH_3–N processed by per unit volume of biological carriers, these coefficients were evaluated as follows: $\beta = 3.72 \times 10^{-13}$, $x_1 = 2.7$, $x_2 = 1$, $x_3 = 1.98$. The value of $R^2 = 84.5\%$, which indicated that the removal rate of COD by biocarriers is strongly correlated with these factors, and the fit of the obtained model was fairly-good. Hence, the removal rate of NH_3–N processed by per unit volume of biological carriers can be expressed as follows:

$$\lambda = 3.72 \times 10^{-13}\theta(1-\varphi)^{2.7}(\rho^2\nu^{1/2}\varepsilon)^{1.98} \tag{6.7}$$

For the removal rate of COD processed by per unit volume of biological carriers, these coefficients were evaluated as follows: $\beta = 4.74 \times 10^{-18}$, $x_1 = 2.64$, $x_2 = 1.12$, $x_3 = 2.94$. The value of $R^2 = 93.2\%$, which indicated that the removal rate of COD by biocarriers was strongly correlated with these factors, and the fit of the obtained model was fairly-good. Hence, the removal rate of COD processed by per unit volume of biological carriers can be expressed as follows:

$$\lambda = 4.74 \times 10^{-18}\theta^{1.12}(1-\varphi)^{2.64}(\rho^2\nu^{1/2}\varepsilon)^{2.94} \tag{6.8}$$

For the removal rate of TP processed by per unit volume of biological carriers, these coefficients were evaluated as follows: $\beta = 1.84 \times 10^{-9}$, $x_1 = 3.4$, $x_2 = 0.65$ $x_3 = 1.2$. But the value of $R^2 = 64.6\%$, which indicated that the fit of the obtained model wasn't fairly-good, which might be due to the low concentration of TP in the inflow MBBR system, and need further study (Fig. 6.11).

6.5 Summary and Conclusion

In this paper, the removal rate of the pollutants in water body by MBBR based on folded cylindrical biological carrier was studied experimentally. Based on the experimental results and the filed test results, the removal effect of MBBR based on folded cylindrical biological carrier on the pollutants in water body was discussed.

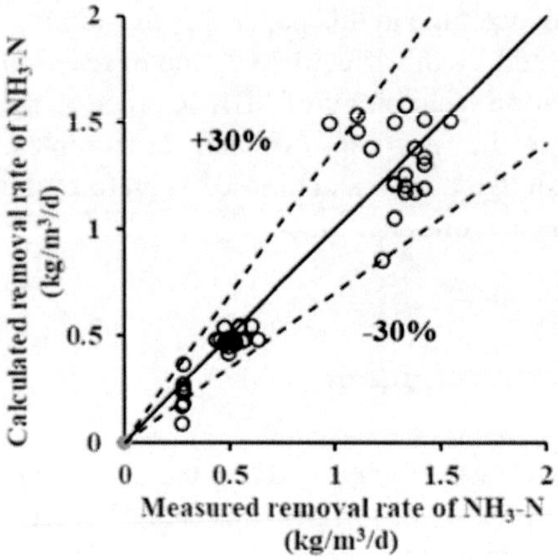

(a) Error analysis of NH_3-N concentration

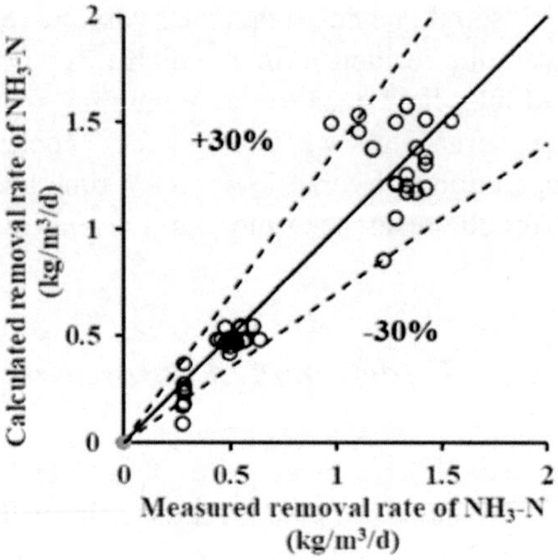

(b)Error analysis of COD concentration

Fig. 6.11 Error analysis between calculated results and measured results of COD and NH_3–N

Further, the calculation model for the removal rate of the pollutants in water body by MBBR based on folded cylindrical biological carrier was derived, which can be used to predict the removal rate of MBBR systems. The main conclusions are as follows.

The MBBR system based on folded cylindrical biological carrier possessed a good treatment effect on the pollutants of COD, BOD, NH_3–N and TP. In addition, the dissolved oxygen concentration of the outflow could be maintained at a high level. The average removal rate of COD and NH_3–N by MBBR system based on folded cylindrical biological carrier reached 95% and 87.75% respectively.

The MBBR system based on folded cylindrical biological carrier possessed a high removal rate of the pollutants in water body. Under the conditions of natural hanging film, the biofilm culture was basically completed on 17th day, which was fast than that of YL-II. Aerobic active biological carrier, K3 biological carrier and polypropylene hollow sphere type of biocarrier carrier.

The calculation model for the removal rate of the pollutants in water bodies by MBBR based on folded cylindrical biological carrier was derived based on the fill rate of reaction tank, the specific surface area of bio-carriers, the sewage density, the kinematic viscosity and the temperature of the water body. The calculation model was a good predictor of the removal rate COD and NH_3–N by MBBR based on folded cylindrical biological carrier, but couldn't be used to predict the removal rate of TP, which might be due to the low content and the variation with a small range of TP in this experiment and the field tests, which need further study.

References

Biswas K, Taylor M, Turner S (2014) Successional development of biofilms in moving bed biofilm reactor (MBBR) systems treating municipal wastewater. Appl Microbiol Biotechnol 98(3):1429–1440

Deng LJ, Guo WS, Ngo HH, Zhang XB, Wang XC, Zhang QH, Chen R (2016) New functional biocarriers for enhancing the performance of a hybrid moving bed biofilm reactor-membrane bioreactor system. Biores Technol 208:87–93

Hu XB, Xu K, Wang Z, Ding LL, Ren HQ (2013) Characteristics of biofilm attaching to carriers in moving bed biofilm reactor used to treat vitamin C wastewater. Scanning 35(5):1–9

Huang Y, Huang C, Liu C, Chen J, Cao J, Shen XC, Huang DH (2022) Application of MBBR process in pharmaceutical wastewater. Water Supply Drain 48(3):75–79

Iordache O, Moga IC, Mitran EC, Sandulache I (2020) BOD & COD reduction from textile wastewater using bio-augmented HDPE carriers. In: The 8th international conference on advanced materials and systems

Leyva-Diaz JC, Martin-Pascual J (2015) Start-up of membrane bioreactor and hybrid moving bed biofilm reactor–membrane bioreactor: kinetic study. Water Sci Technol A J Int Assoc Water Pollut Res 72(11):1948–1953

Leyva-Diaz JC, Gonzalez-Martinez A, Calderon K, González-López J, Muñío MDM, Poyatos MJ (2016a) Microbial kinetics and enzymatic activities in hybrid moving-bed biofilm reactor-membrane bioreactor systems. Chem Eng Technol 39(6):1067–1076

Leyva-Diaz JC, Martin-Pascual J, Poyatos JM (2016b) Moving bed biofilm reactor to treat wastewater. Int J Environ Sci Technol 14(4):1–30

Ma WF, Guo H, Han DM (2014) Research on the effects of modified HDPE for wastewater treatment. Adv Mater Res 1010–1012:532–536

Mahdavi H, Norouzian S (2018) Preparation and characterization of modified ultra-filtration nylon 6 membrane modified by poly (acrylamide-co-maleic anhydride). J Polym Res 25(10):1–12

Mczaowski A (2019) Analysis and evaluation of the measurement system based on kama orifices for measuring the flow of sewage in the light of the water law requirements. Acta Scientiarum Polonorum Formatio Circumiectus 2(2):147–157

Ogunsona E, Misra M, Mohanty A (2017) Accelerated hydrothermal aging of biocarbon reinforced nylon biocomposites. Polym Degrad Stabil 139:76–88

Peng Z, Wu H, Ding M, Li M, Huang X, Zheng R, Xu L (2021) Ecological compensation standard of a water-receiving area in an inter-basin water diversion based on ecosystem service value and public willingness: a case study of Beijing. Sustainability 13(9):1–15

Rodriguez-Sanchez A, Leyva-Diaz C, Gonzalez-Lopez J, Poyatos J (2018) M: Membrane bioreactor and hybrid moving bed biofilm reactor-membrane bioreactor for the treatment of variable salinity wastewater: influence of biomass concentration and hydraulic retention time. Chem Eng J 336:102–111

Schopf A, Delatolla R, Mathew R, Tsitouras A, Kirkwood KM (2018) Investigation of copper inhibition of nitrifying moving bed biofilm (MBBR) reactors during long term operations. Bioprocess Biosyst Eng 41(10):1485–1495

Shuvalov MV (2020) System analysis of normative requirements evolution in terms of composition and characteristics of sewage discharged into surface water bodies. Urban Constr Architect 10(2):43–56

Sun ZJ, Tong Y, G W, Zhu J, Chen Y (2021) Application of MBBR Process based on modified filler for high-load treatment of starch sugar wastewater. Bioprocess Process 19(3):288–293

Tang JW, Ning K, Lei WX, Zhang CH, Meng XF (2018) Treatment of acrylic fiber wastewater by Fenton fluidized bed oxidation process with suspended filler coated iron oxide. Chem Ind Eng Progr 37(11):4476–4484

Tsitouras A, Butcher J, Li J, Stintzi A, Delatolla R (2022) Biofilm morphology and microbiome of sequencing batch moving bed biofilm reactors treating cheese production wastewater. Bioresour Technol Rep 17:100898

Wang JH (2018) Manufacturing technology of conformal cooling plastic mold based on 3D printing technology. Die Mould Manuf 18(01):71–76

Wang K, Ke SZ, Yuan HZ, Zhu J, Li JW (2020) Effect of ammonia nitrogen concentration on microbial community structure in MBBR process. Environ Eng 38(9):119–125

Wang R, Wei B, Zhu J, Zhao XD, Yu GL (2022) Study on the hydrodynamic performance and treatment effect of a modified biological carrier in wastewater treatment. Sci Total Environ 844:156974

Zhang XB, Zhang ZM, Liu Y, Ngo HH, Guo WS, Wang HZ, Zhang YF, Zhang D (2020) Impacts of sulfadiazine on the performance and membrane fouling of a hybrid moving bed biofilm reactor-membrane bioreactor system at different C/N ratios. Biores Technol 318:124180

Chapter 7
Salt-Water Intrusion Prediction Using an Artificial Neural Network in the Bouregreg Estuary (Morocco)

S. Haddout, K. L. Priya, A. M. Hoguane, and I. Ljubenkov

Abstract Various numerical models have been used to describe hydrodynamics and predict salinity. Whenever salinity is predicted or when parametric data is unavailable or poorly estimated, these equations are even more complicated. Salinity simulation studies are therefore conducted using Artificial Neural Networks (ANN) based on back-propagation neural networks (BPNNs). In this short paper, a simple technique adopting an artificial neural network (ANN) for predicting the salinity data in the Bouregreg estuary (Morocco) has been attempted. As a result, it can be used directly to assess salinity parameters in the Bouregreg estuary because it is transparent and applicable to the estuary. The Nash–Sutcliffe Efficiency Coefficient (NSC), Root Mean Squared Error (RMSE), and Normalized Objective Function (NOF) are used to assess the performance of the artificial neural network model. The research results showed that the ANN model achieved high salinity prediction performance, with the small RMSE coefficient, the NOF function being less than one, and the NSC coefficient being very close to 1. In terms of water resource development in this area and support for future management plans, this forecasting algorithm is a good example of an effective tool.

Keywords Bouregreg Estuary (Morocco) · Salt-intrusion · Statistical parameters · Artificial neural network

S. Haddout (✉)
Faculty of Science, Department of Physics, Ibn Tofail University, B.P. 133 Kenitra, Morocco
e-mail: soufian.haddout@gmail.com

K. L. Priya
Department of Civil Engineering, TKM College of Engineering, Kollam 691005, India

A. M. Hoguane
Centre for Marine Research and Technology, Eduardo Mondlane University, PO Box 128, Quelimane, Mozambique

I. Ljubenkov
Water Development Ltd., Kvaternikova 7, Split, Croatia

H. Xu (ed.), *Proceedings of the 5th International Symposium on Water Resource and Environmental Management*, Environmental Science and Engineering,
https://doi.org/10.1007/978-3-031-31289-2_7

7.1 Introduction

Estuaries play an essential role in the human-earth system by connecting river and ocean waters and where the mixing of ocean and river water affects the health of aquatic ecosystems (Savenije 2015). This makes the functioning of estuary systems an important area of research. A crucial element of estuarine dynamics is the interaction between salt and fresh water (Haddout and Maslouhi 2017; Haddout et al. 2017). The intrusion of salt into water is a major cause of quality issues for industrial and agricultural uses as well as stratification and anoxia of bottom waters due to the intrusion of salt (Kurup et al. 1998; Etemad-Shahidi et al. 2011) and sedimentation of rivers due to saline intrusion. Both engineers and scientists are interested in the effects of human salt intrusions. Hydrodynamics and salt intrusion have been described using numerical models extensively. Each case must be calibrated and verified before a numerical model can be applied. The equations involved are even more complicated when there is scarce field data for model validation. A relatively new computational approach, artificial neural networks (ANNs), offers an alternative method to understand and manage hydrodynamic processes one step ahead, as numerical models have difficulties and challenges in predicting salinity variations. Artificial neural networks are widely accepted in many disciplines and offer an alternative method to predict salinity fluctuations. A number of properties, including nonlinearity, parallelism, noise tolerance, and ability to learn, make ANNs a good choice for this application (Hagan et al. 1995; Wu et al. 2009). In this study, a backpropagation neural network model for time-series salinity prediction in the Bouyregreg Estuary (Morocco) was implemented with input forcing functions affecting water level and freshwater discharge.

7.2 Materials and Methods

7.2.1 Study Site

Along the Atlantic coast, between Sale and Rabat, the Bouregreg estuary is situated at 34°N and 6°50′W (Fig. 7.1). The Sidi Mohammed Ben Abdellah dam restricts the river's maximum length and width to 23 km. The average freshwater flow ranged from 3–84 m^3s^{-1}. There is also a semi-diurnal tidal cycle of 12.25 h at the mouth of the estuary, which is characterized by a narrow tidal range of 2.3 m. It is primarily composed of seawater, so there is little freshwater influx into this estuary most of the year (Haddout 2020). The estuary's saltwater intrusion has gotten worse over time since the Sidi Mohammed Ben Abdellah Dam was constructed.

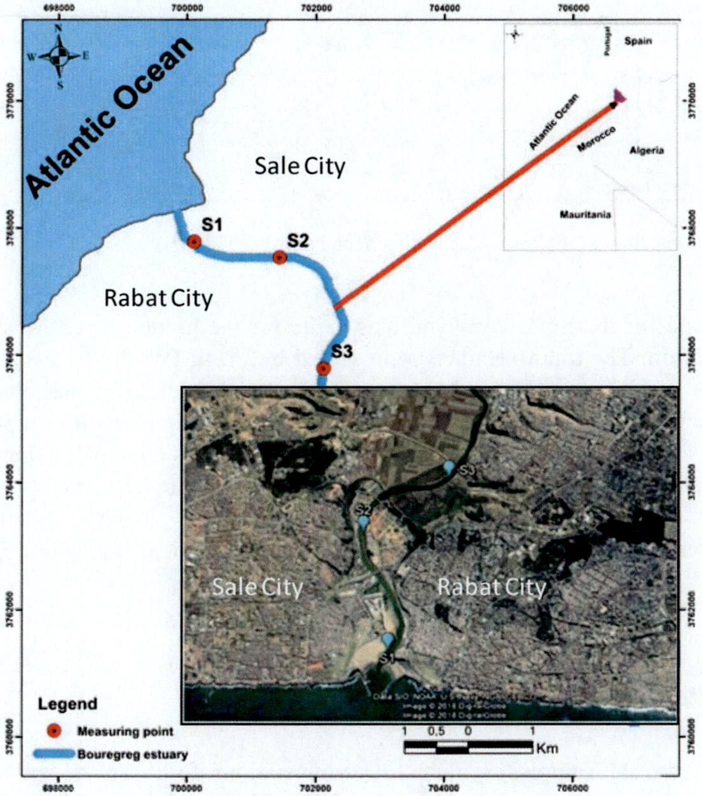

Fig. 7.1 Bouregreg estuary and location of measuring sites

7.2.2 Artificial Neural Networks

The neurons and nodes that make up a neural network are simple processing elements. Direct communication links connect each neuron to other neurons, each with its own weight. In order to solve a problem, the network needs to know the weights that represent the information. The most commonly used perceptron, called a perceptron with a hidden layer, is composed of an input, a hidden layer and an output layer. There are many types of neural networks depending on their structure. The most popular ANN is the back-propagation neural network (BPNN), which was developed by Rumelhart et al. (1986) and is used in the current study. It can be used to address a variety of nonlinear issues because it is a multiple-layer network with nonlinear differentiable transfer functions. BPNNs are trained using input vectors and corresponding target vectors until they can approximate a defined minimum error or a specified maximum number of epochs. Any function that contains a finite number of discontinuities can be approximated by BPNNs that have biases, weightings, sigmoid layers, and linear output layers. According to Fig. 7.2, the BPNN has the following architecture. A

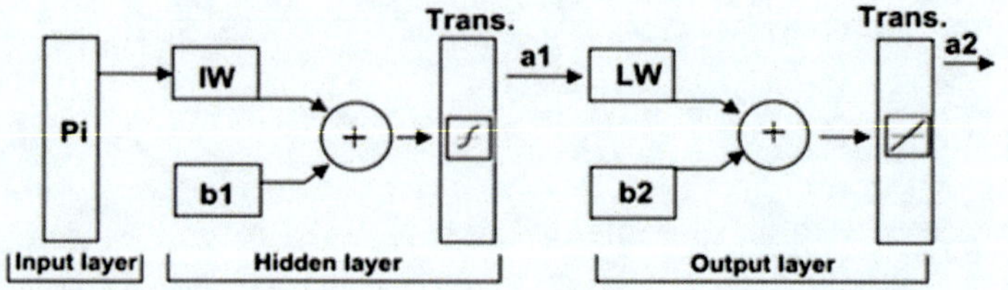

Fig. 7.2 Architecture of the back-propagation neural network (BPNN)

weight value for the input layer and bias value for the hidden layer are shown in the illustration. The input vector is represented by Pi. In IW and b1, we show the input layer weights and the hidden layer biases. The hidden layer bias and weight are represented by b2 and LW in this illustration. The input vector is represented by Pi. As sigmoid and linear transfer functions can extrapolate beyond the range of training data, they are widely used in the hidden layer and output layer.

Matlab was used to implement the Artificial Neural Network model, which is based on the Levenberg–Marquardt technique included in the Neural Network Toolbox.

7.2.3 Data Observation

We collected water samples from the following stations during spring and neap tide on October 8th, 2017 and February 16th, 2017: High and low tide samples were collected during the slack period. Keeping pace with the tidal wave, the estuary was measured upstream. The National Agency of Port (Morocco) provided the water level at station 1. The freshwater discharge data at station 3 has been obtained. Salinity is measured in-situ using an YSI Pro30 conductivity meter connected to a 10 m cable. The positions of each measurement were also recorded using a Global Positioning System (GPS). Various measurements can be carried out at low cost using the devices used.

7.2.4 Evaluation Criteria

BPNN's performance in predicting salinity in the Bouregreg estuary was assessed by analyzing the correlation between predicted and observed salinity at station 2. An analysis of residual errors was conducted based on the Mean Square Error (RMSE), the Normalized Objective Function (NOF), and the Nash–Sutcliffe (NSC) coefficient [more information can be found in Haddout et al. (2020).

7.3 Results and Discussion

The hourly salinity measurement data at station 2 were collected for calibration and validation phases using the ANN model (BPNN). This work creates a BPNN model to predict estuary salinity at station 2 using water-level data at station 1 and freshwater discharge data at station 3. Figures 7.3 and 7.4 show the time-series of water level at station 1 (downstream) and the time-series of hourly river freshwater discharges at station 3 (upstream).

Up until the perfect architecture was found, the BPNN model was calibrated. The optimum BPNN architecture was determined to be 2-4-1, which consists of input neurons, hidden neurons, and output neurons (see Fig. 7.5).

According to Fig. 7.6 the observed salinities at station 2 were compared with BPNN predicted salinities during calibration and validation. During the calibration and validation phase, it was found that the BPNN model overestimated the minimum salinities during ebb tide and underestimated the peak during flood tide. RMSE, NOF, and NSC were 0.88 ppt, 0.052, and 0.87, respectively, during the calibration, while they were 0.75 ppt, 0.034 ppt, and 0.76 during the validation phase, as shown in Table 7.1.

Based on the BPNN model, Fig. 7.7 shows the comparison between measured and predicted salinity. A best-fit line was found to produce salinities evenly distributed around it. According to the results, the ANN (BPNN) approach reproduces non-linear relationships between input and output. In addition to predicting salinity variations,

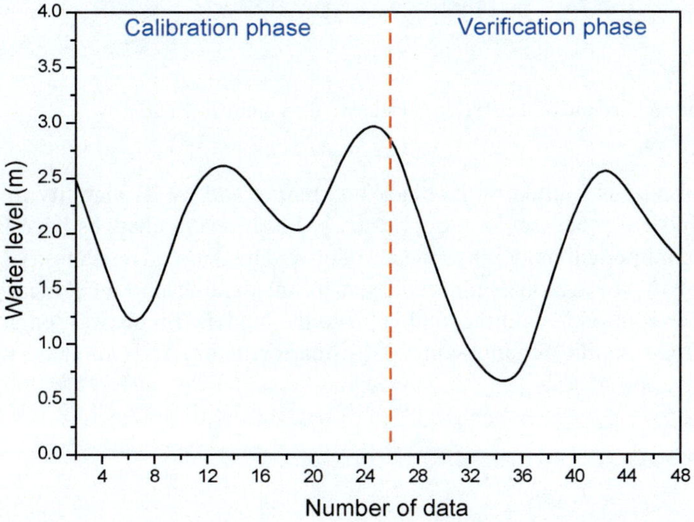

Fig. 7.3 Time series water level at station 1 used for the calibration and validation phases (x-axis is in hours)

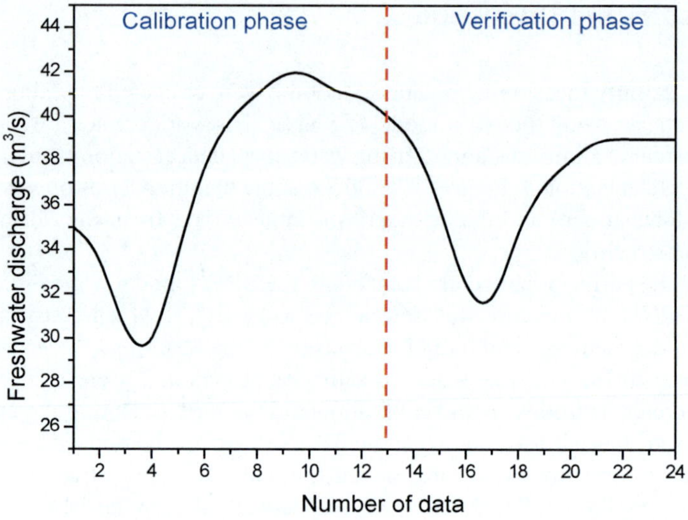

Fig. 7.4 Upstream river freshwater discharges at station 3 used for the calibration and validation phases (x-axis is in hours)

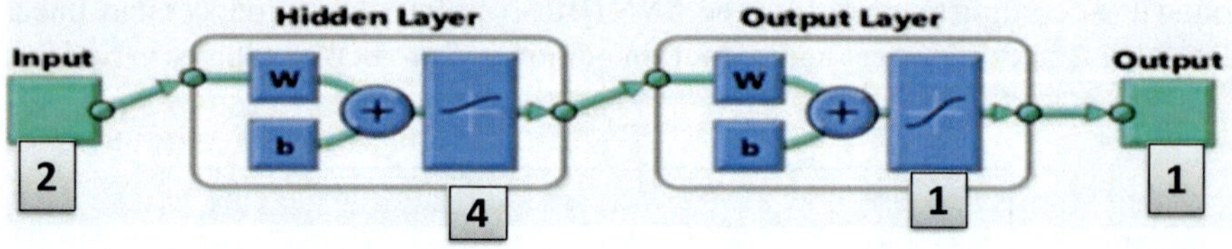

Fig. 7.5 The architecture of the BPNN-model used for salinity modelling

the ANN model is limited by its black box nature and by its inability to simulate internal physical processes in a tidal estuary. Engineering changes have led to the adoption of numerical models as a means of providing detailed resolutions. However, before applying these models for prediction, a considerable amount of measurements data will be required to calibrate and validate the models. In this way, engineers can quickly assess salinity changes in tidal estuaries using ANN models, which are data-driven approaches.

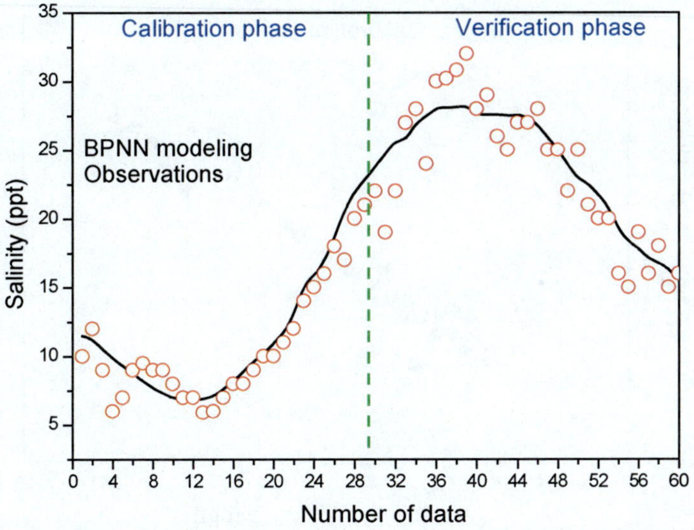

Fig. 7.6 The variation of the measured and modeled using BPNN model salinities at the station 2 during the calibration and validation. Note that red circles are measurements and the solid line is the BPNN technique (x-axis is in hours)

Table 7.1 Results of the predictive model (BPNN) of salinity at Station 2

Statistical analyses	RMSE (ppt)	NOF (–)	NSC (–)
St.2 calibration	0.88	0.052	0.87
St.2 validation	0.75	0.034	0.76

7.4 Conclusions

Salinity is an essential parameter for assessing the condition of the estuarine aquatic ecosystem. The ANN (BPNN) model was used in the current study to predict the salinities of the estuaries. A driving force function is constructed through the use of time series of freshwater river discharges as input and the time series of water levels upstream and downstream. Although the BPNN technique is a black-box model, the salinity values estimated by the model are good. A large and suitable amount of input–output data sets would further increase predictability of the ANN model since it is a data-driven technique.

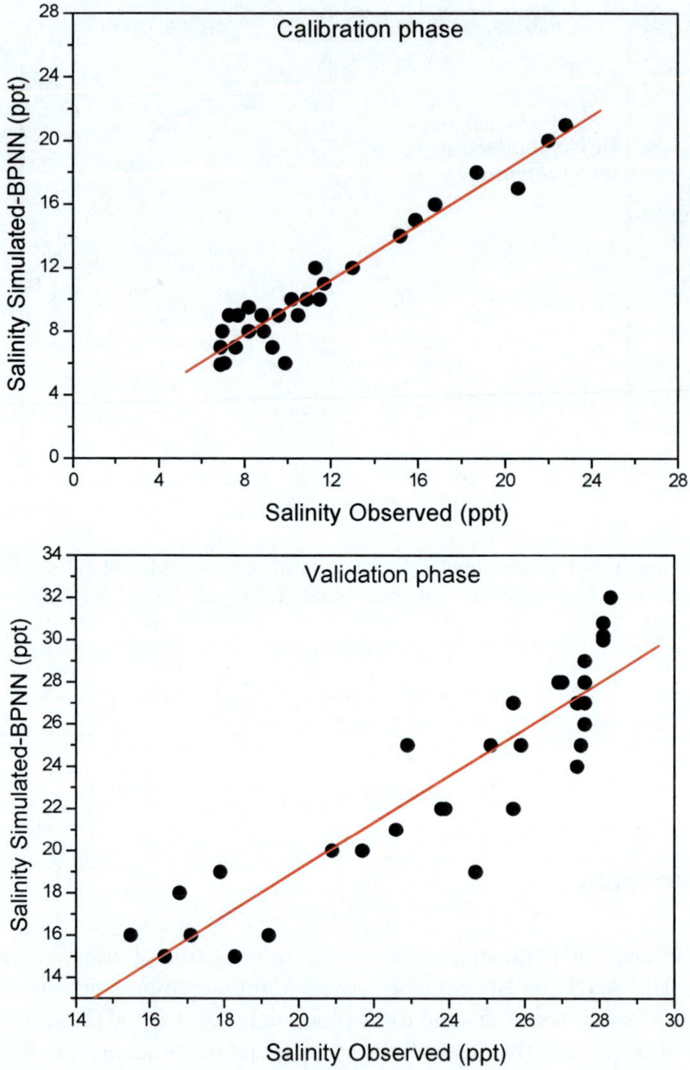

Fig. 7.7 Measurements versus modelling using BPNN approach

Acknowledgements In appreciation of two anonymous reviewers, the authors acknowledge their scientific suggestions and constructive comments which helped improve the manuscript significantly.

References

Etemad-Shahidi A, Parsa J, Hajiani M (2011) Salinity intrusion length: comparison of different approaches. Proc Inst Civ Eng-Marit Eng 164:33–42

Haddout S (2020) A power-law multivariable regression equation for salt intrusion length in the Bouregreg estuary, Morocco. Mar Georesour Geotechnol 38(4):417–422

Haddout S, Maslouhi A (2017) Two-dimensional modeling of the vertical circulation of salt intrusion in the Sebou estuary under different hydrological conditions. ISH J Hydraul Eng 1–18

Haddout S, Igouzal M, Maslouhi A (2017) Modeling the effect of salt water intrusion in the Sebou River estuary (Morocco). Russ Meteorol Hydrol 42(12):803–811

Haddout S, Priya KL, Adarsh S (2020) A predictive model for salt intrusion in estuaries applied to the Muthupet estuary (India) and Bouregreg estuary (Morocco). ISH J Hydraulic Eng 26(4):430–447

Hagan MT, Demuth H, Beale M (1995) Neural network design. PWS Publishing Company, Boston

Kurup GR, Hamilton DP, Patterson JC (1998) Modelling the effect of seasonal flow variations on the position of salt wedge in a microtidal estuary. Estuar Coast Shelf Sci 47(2):191–208

Rumelhart DE, Hinton E, Williams J (1986) Learning internal representation by error propagation. Parallel Distrib Process 1:318–362

Savenije HHG (2015) Prediction in ungauged estuaries: an integrated theory. Water Resour Res 51(4):2464–2476

Wu CL, Chau KW, Li YS (2009) Methods to improve neural network performance in daily flows prediction. J Hydrol 372(1–4):80–93

Chapter 8
Progress and Foresight in Wetland Environmental Science Research: An Econometric Analysis Based on the Knowledge Map of the WOS Core Database Literature

Feiyan Huang, Yong Zhang, and Yunyao Ma

Abstract The wetland environment faces the risk of degradation or disappearance, which will drive the demand for scientific research on the wetland environment. This paper uses CiteSpace software based on the bibliometric method to visually analyze the relevant works of literature on wetland environmental science in the core database of the Web of Science from 1977 to 2021. The results showed that the number of literature related to wetland environmental science increased rapidly after 2005. The United States has the most significant influence in this field, with a total of 1,892 articles published. Joan García, Baoshan Cui, Junhong Bai, and other authors made outstanding contributions. The emergent analysis of keywords showed that the hot spots of research in the last five years were the effects of human activities or natural factors on carbon storage in wetlands, the effects of wetland pollution on microbial communities, and the degradation and restoration of wetlands. Cluster analysis yielded 23 valid clusters, of which "pattern", "Atmospheric methane", and "Mercury" were the three most significant clusters. According to the content of cluster labels, they could be divided into two categories: the relationship between wetlands and climate change and wetland pollution and control. Future research on wetland environmental science may focus on wetland classification, key influencing factors of wetland ecosystems, the combination of wetland bioremediation technology with nanotechnology or transgenic technology, and wetland restoration measures.

Keywords Wetlands · Environmental science · CiteSpace · Bibliometrics · Literature summary · Research progress

F. Huang · Y. Zhang (✉) · Y. Ma
School of Earth Sciences and Spatial Information Engineering, Hunan University of Science and Technology, Xiangtan 411201, Hunan Province, China
e-mail: 292278@QQ.com

H. Xu (ed.), *Proceedings of the 5th International Symposium on Water Resource and Environmental Management*, Environmental Science and Engineering,
https://doi.org/10.1007/978-3-031-31289-2_8

8.1 Introduction

Wetlands have been widely concerned because of their essential environmental regulation function and ecological benefits (Wang et al. 2021). Since the twentieth century, with the rapid development of the social economy and the intensification of human activities, the problem of wetland environmental degradation has become increasingly prominent. By 2009, at least 33% of the world's wetlands had disappeared (Hu et al. 2017). The rapid reduction of wetlands will directly destroy the normal function of the wetland environment and will cause severe global ecological environment problems, which will threaten human health and development. It is vital to carry out scientific research on the wetland environment and make clear the future development direction.

With the official adoption of the Convention on Wetlands in 1971, a series of studies have been conducted worldwide on climate change, pollution status, and wetland restoration. Research on wetlands and climate change mainly predicts the interaction between wetlands and climate change through models. Robert et al. established a nonlinear dynamic model and predicted that by 2080, 22% of the world's coastal wetlands will disappear due to sea level rise alone (He et al. 2014). The primary pollutants in the wetland are heavy metals and organic pollutants (Li et al. 2022). Bai et al. found that the level of iron and chromium pollution in the soil of Wanqingsha wetland in the Pearl River Estuary had reached a severe level (Bai et al. 2011). Research on wetland restoration has focused on assessment methods, techniques, and success indicators. With the help of ecosystem, category, and variable indices, Staszak et al. constructed an index of ecosystem integrity to evaluate the success rate of salt marsh wetland restoration (Staszak and Armitage 2013). At the same time, some pieces of literature have reviewed the research on wetland carbon budget, heavy metal pollution, vegetation methane transport, and constructed wetlands (Ravikumar et al. 2022). Although these reviews have enhanced the knowledge of wetlands, these studies tend to focus only on a particular type or region of wetlands. Therefore, it is essential to systematically review the development of wetland environmental science in the past decades and to clarify the general direction of wetland environmental science research in the future.

Bibliometrics takes the characteristics of literature as the research object. With the visualization function of bibliometrics software, researchers can objectively evaluate the historical evolution process of the research field in a certain period (Wu et al. 2021). The CiteSpace software is one of the most widely used bibliometrics software, which can intuitively display the position and size of nodes in the knowledge network and quickly lock the most critical and core information in the field (Zhu et al. 2018). The Web of Science (WOS) core collection connects multiple regional citation indexes. WoS can quickly locate the most cutting-edge and classic literature in a specific research field.

This paper summarizes global wetland environmental scientific research's development process and future direction. Based on the WoS Core Collection database, the literature published on wetlands from 1977/1/1 to 2021/1/1 was screened. This paper uses the bibliometric software CiteSpace to analyze from the perspectives of time evolution, cooperative relationship network, and keyword analysis. This study will well summarize the development trend of global wetland environmental science and provide a reference for the future development direction of this field.

8.2 Methodology

8.2.1 Data Collection and Cleaning

Search the WOS Core Collection database with the subject "Wetland", and select "Environmental Sciences" as the research category and "Article" as the literature category. The search was limited to January 1, 1977, to December 31, 2021. In order to select high-quality literature, 23 journals belonging to Q1 journals in the JCR journal division for 2020–2021 were selected from the top 100 journals in the above search results by using the "source publication name" function in WOS (Table 8.1). At the same time, the retrieval function of WOS was used to exclude conference-related papers. The manual screening of the results was carried out to eliminate the review and irrelevant literature to ensure the data's reliability and validity. Finally, 5952 pieces of literature related to wetland environmental scientific research were obtained. The selected pieces of literature are downloaded and saved as TXT files in WOS in the format of "Full Record and Cited References" as samples for data analysis.

8.2.2 Data Analysis

This paper mainly uses the WoS data analysis section of CiteSpace (version 5.8R3) to conduct statistical analysis and visual analysis of the number of publications, countries, authors, and research keywords on wetland environmental science. CiteSpace parameters can be set as follows: Time threshold: 1977–2021; Node types: Country, Author, Keyword; To make the bibliometric mapping clearer and more explicit, the authors chose Pathfinder, Pruning sliced networks, and Pruning the merged network for the pruning method; the rest of the options are default.

Table 8.1 Journal information

Serial number	Publication titles	Impact factor 2020–2021
1	Science of the Total Environment	7.963
2	Water Research	11.236
3	Environmental Pollution	8.071
4	Biological Conservation	5.990
5	Water Resources Research	5.240
6	Global Change Biology	10.863
7	Journal of Hazardous Materials	10.588
8	Global Biogeochemical Cycles	5.703
9	Conservation Biology	6.560
10	Journal of Environmental Management	6.789
11	Chemosphere	7.086
12	Journal of Environmental Sciences	5.565
13	Ecotoxicology and Environmental Safety	6.291
14	Environmental Research Letters	6.793
15	Marine Pollution Bulletin	5.553
16	Remote Sensing of Environment	10.164
17	Ecosystem Services	5.454
18	Environmental Modelling Software	5.288
19	Environmental Science Policy	5.581
20	Atmospheric Chemistry and Physics	6.133
21	Environmental and Experimental Botany	5.545
22	Environmental Research	6.498
23	Environmental Technology Innovation	5.263

8.3 Results

8.3.1 Literature Panorama

Literature quantity analysis. Since 1977, the number of publications on wetland environmental science has been increasing. In the past 46 years, the total number of publications can be divided into three stages: 1977–1988 was the early exploration stage, during which the annual average number of publications was less than 10. The period from 1989 to 2004 was a slow growth stage. In 2004, the number of articles exceeded 100 for the first time. Since 2005, the number of published papers has soared, reaching 822 in 2021, 6.5 times that in 2005 (Fig. 8.1), reflecting that the research of wetland environmental science is gaining wide attention.

The number of national publications. The number of national publications can reflect the importance and influence of the country on the research field. In wetland environmental science, the top 10 countries in terms of publication volume are the United States, China, Canada, Australia, England, Germany, Spain, France, the Netherlands, and Italy. The United States has a long wetland environmental science research history, which began in the 1970s. Around 1970, the United States established the Environmental Protection Agency (EPA) and issued the National Environmental Policy Act of 1969 to promote the research of wetland environmental science. China's research on wetland environmental science started relatively late, at the end of the 1990s. However, with the rapid development of the Chinese economy and the improvement of scientific research, wetland environmental research has been paid more and more attention. In particular, in the past five years, China's publication

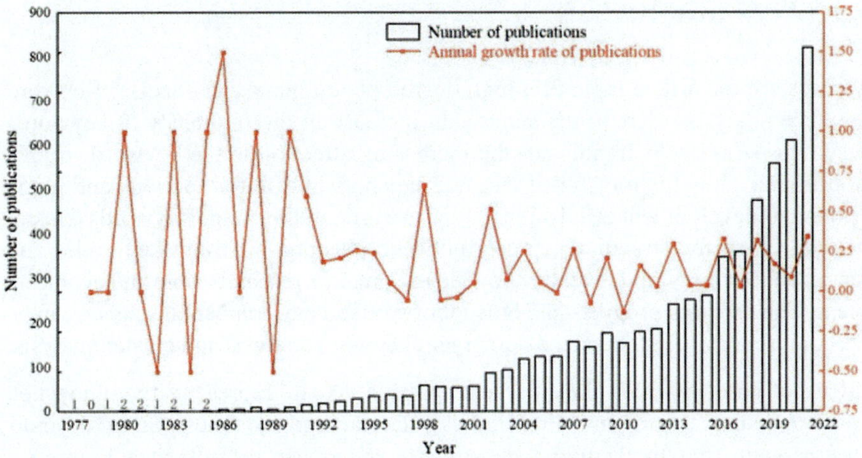

Fig. 8.1 Time evolution of publications

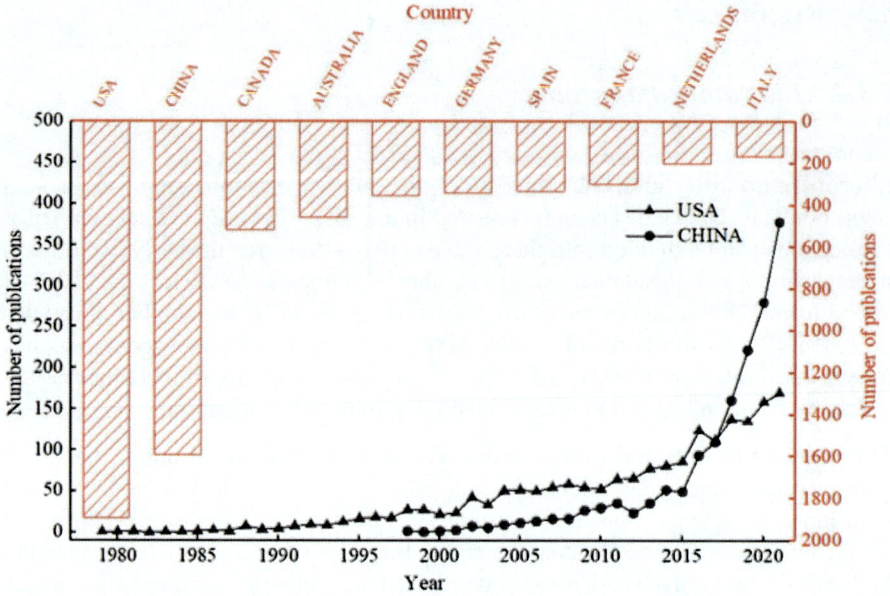

Fig. 8.2 Time evolution of the number of national publications and the number of publications by China and the United States

volume in this field has shown a rapid growth trend (Fig. 8.2), achieved fruitful results, and gained increasing global influence.

8.3.2 *Keywords Analysis*

Keywords are the article topic of a high degree of summary and concise. Keyword emergence analysis refers to the significant increase in the frequency of keywords in a certain period, which indicates the increasing attention to the keywords in this period. It can show the transfer of research hotspots in different periods and judge the potential development trend and frontier research in this field. Keywords Cluster view can reflect the structural features between clusters, highlight key nodes and essential connections, and reveal a knowledge domain's research topic and evolution process. The research progress and hotspot of wetland environmental science can be more comprehensively reflected by emergent keyword analysis and cluster analysis.

Analysis of emergence. In the emergent analysis, "Year" represents the time when the node appears; "Begin" and "End" indicate the start and end times. "Strength" indicates emergent strength; the higher the strength, the greater the influence. Figure 8.3 shows the top 25 most cited breakout keywords. It can be divided into three stages according to the change of emergent time. Stage 1: 1987–2006, "atmospheric methane", "carbon dioxide", and "flux" were the first to appear. Combined Fig. 8.2

shows that the study of the wetland environment in this period is on the right track. Currently, the research focuses on quantitatively analyzing wetland methane, carbon dioxide, and other greenhouse gases (Turetsky et al. 2014). Stage 2: 1992–2011, oxidation, cadmium, habitat, methyl mercury, reed bed, and other keywords appear. At this stage, the characteristics of heavy metal pollution, the effects of organisms on heavy metals (Bonanno and Giudice 2010), the interaction between climate change and wetland ecosystem, and the effects of alien species or grazing on wetland ecological environment are the hot spots of attention. It is worth noting that the presence of keywords methane emission and emission indicates that the sources and sinks of greenhouse gases such as methane in wetlands are still the focus of research. Stage 3: 2017–2021, the keywords bacterial community, microbial community, and blue carbon began to appear. It reflects that the research focus in recent years is the impact of human or natural activities (climate change, sea level rise) on wetland carbon storage (Adhikari et al. 2019), the impact of pollutants on wetland microbial community (Worthy et al. 2000), and the degree of wetland degradation and restoration measures.

Cluster analysis. In cluster analysis, the Log-likelihood ratio (LLR) algorithm is used based on the keyword co-occurrence network, so words with larger LLR are

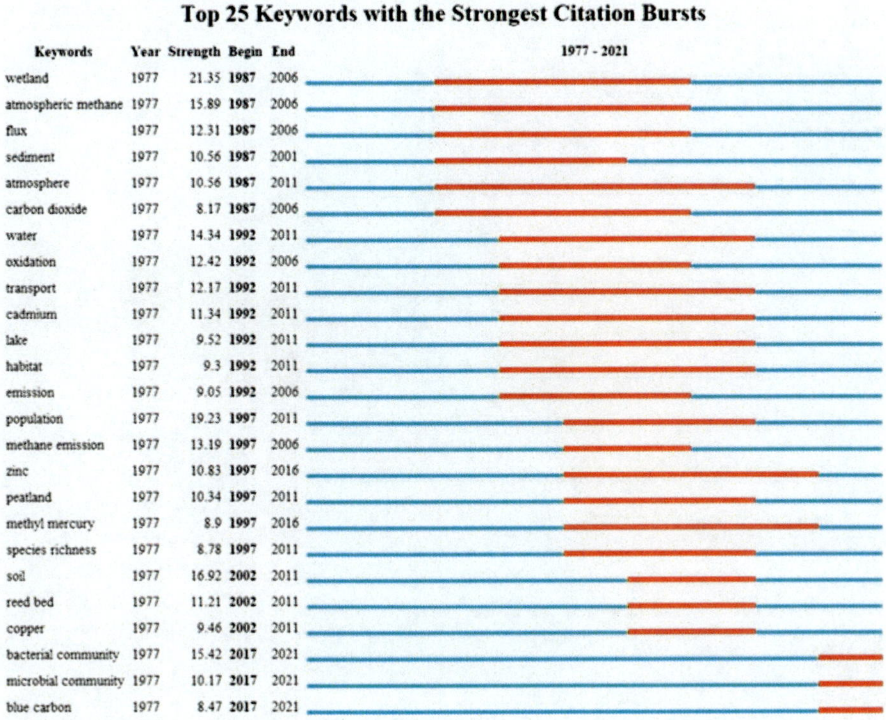

Fig. 8.3 Top 25 keywords with the strongest citation bursts

more likely to represent the cluster. The clustering display Modularity Q value is 0.8436, and the clustering Mean Silhouette value is 0.9358 (Fig. 8.4). When $Q > 0.3$, it means that the community structure is significant; Analysis results are considered convincing when $S > 0.7$ (Zhao and Qin 2022). The results show that the cluster graph is a significant and convincing community structure. Finally, 23 effective clusters were obtained, and cluster numbers ranged from #0 to #23. The smaller the number, the larger the scale of literature research under the cluster. The label can be classified into two categories based on the label content.

The relationship between wetlands and climate change: The clusters of "Pattern" (#0), "Atmospheric methane" (#1), "Methane" (#6), "Fluxe" (#10), and "Emission" (#21) all reflect the impact of wetlands on climate change. Due to their unique conditions, wetlands can slow down the decomposition of CO_2 and generate net carbon sinks, and may also increase CH4 emissions, thus affecting the global climate (Song et al. 2009). Mohamed et al. found that wetland drainage has a significant impact on microclimate. In the dry season, the relative humidity will decrease by 30–40%, and the temperature will rise by 4–6 °C (Mohamed et al. 2005). The clustering of the Salt marsh (#3), Climate change (#7), DOC (#14), and Dissolved organic matter (#17) reflects the impact of climate change on wetlands. The research of Watson, Ivan et al. showed that the rise of sea level would affect the distribution of wetland vegetation structure (Mohamed et al. 2005).

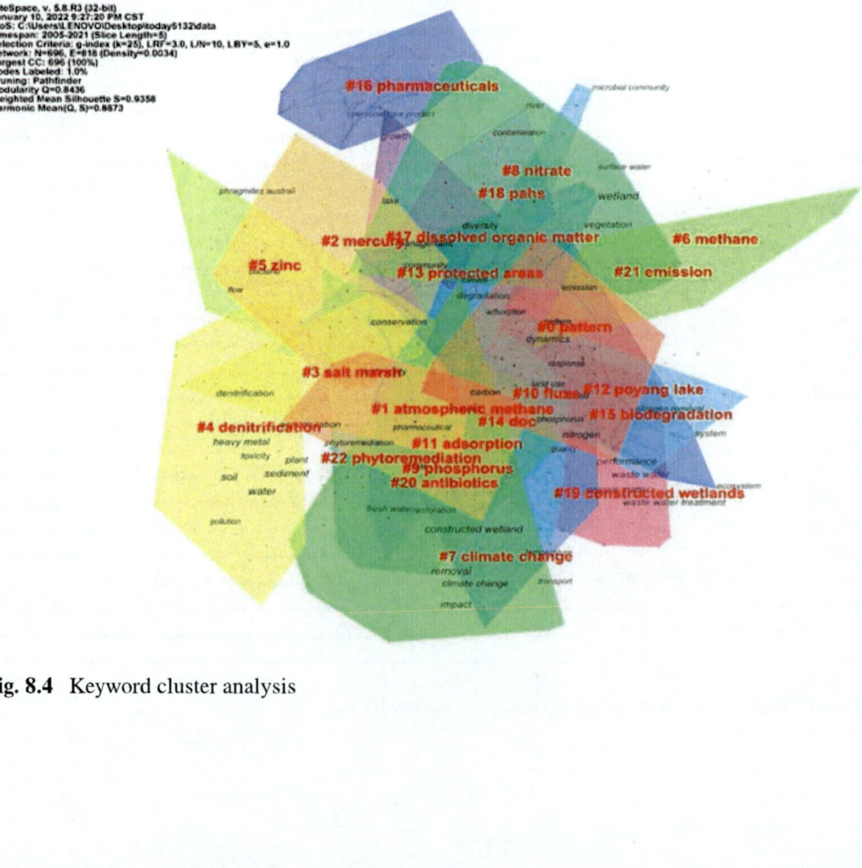

Fig. 8.4 Keyword cluster analysis

Wetland pollution and treatment: The cluster of "Mercury (#2)", "Zinc (#5)", "Protected areas" (#13), "Dissolved organic matter" (#17), "Pahs" (#18) shows people's attention to heavy metals, organic pollutants, and treatment methods in wetlands. Chen et al. added biochar to soil and found that the accumulation of Cd, Pb, and Cu in plant tissues decreased by more than 25% on average (Chen et al. 2018); Liang et al. showed that biochar and compost also had a significant impact on the availability and morphology of heavy metals in wetlands (Liang et al. 2017). The cluster, including "Denitrification" (#4), "Nitrate" (#8), "Phosphorus" (#9), and "Constructed wetlands" (#19), mainly research the performance of constructed wetlands in removing nutrients and emerging organic matter. Gao et al. developed a new type of electrolytic-integrated horizontal subsurface flow-constructed wetland, which can remove more than 80% of nitrate and phosphorus (Gao et al. 2017). The removal rate of emerging organic pollutants by the conventional activated sludge process is minimal. Studies have shown that constructed wetlands effectively remove such pollutants. Yi et al. removed persistent organic pollutants and endocrine-disrupting chemicals through composite-constructed wetlands, with a removal rate of more than 90% (Yi et al. 2017).

8.4 Discussion

Through the analysis results of keywords emergence and clustering, it can be found that the relationship between wetlands and climate change, as well as wetland pollution and control throughout the scientific research of wetland environment, but there are still many problems worth exploring in the process of research. For example, in the quantitative study of wetland greenhouse gases, the degree of wetland degradation, and the influencing factors of wetland pollutants, the results obtained by different researchers are quite different. Therefore, this paper will discuss the relationship between wetlands and climate change, as well as the future research direction of wetland pollution and treatment, in combination with the frontier technology related to environmental science or other disciplines, to provide a reference for the development of wetland environmental science.

8.4.1 The Relationship Between Wetlands and Climate Change

Based on the above results, the development of the relationship between wetlands and climate change can be thoroughly interpreted. Since the late 1990s, urbanization has intensified the degradation of global wetlands, raising concerns about the interaction between wetlands and the atmosphere (Hu et al. 2017; He et al. 2014). It is widely believed that wetlands may affect climate change as carbon sinks. Pickett-Heaps et al.

(2011) quantified methane emissions from Hudson Bay lowland wetlands using a chemical transport model, but the estimated amount of this model was much higher than the research results of Worthy et al. (2000). Song et al. studied for the first time the exchange of CH_4, N_2O, and CO_2 between three wetland ecosystems and the atmosphere in the Sanjiang Plain of China, showing that the gas fluxes between different wetland types are very different (Song et al. 2009). The above studies show that the current research on greenhouse gases in wetlands is still challenging. The main influencing factors must be considered more comprehensively and integrated into the model. The type of wetland and the circulation of greenhouse gases within the wetland ecosystem also need to be considered. At the same time, the impact of climate change on wetlands cannot be ignored. Bohn et al. combined the three models of hydrology, ecology, and emission, and found that under the combined action of temperature rise and precipitation increase, the annual methane emission of wetlands would be doubled (Bohn et al. 2007). Baldwin et al. (2014) and Jennifer et al. simulated climate warming or ozone concentration increase in the laboratory. The results showed that wetland community biomass and species richness would be affected. Most of the above research results are based on models or laboratory cultures, making it difficult to accurately assess the extent of the impact of wetlands on climate change.

In recent years, it has been one of the research hotspots to evaluate the protection degree of wetlands by various protective measures under the influence of climate. Kuhfuss et al., based on a worst-case scenario of a one-meter sea level rise by 2100, which places the area at risk of inundation, showed that post-construction measures could have inundation losses (Kuhfuss et al. 2016). The studies are also evaluated based on different assumptions in which the ecosystem does not change. When it comes to the distant future, ecological services may have changed. Therefore, in the future study of the relationship between wetlands and climate change, it is the key to improve the accuracy of assessment to make a more detailed classification of global natural wetlands and explore the determinants of each wetland type ecosystem.

8.4.2 Wetland Pollution and Treatment

Heavy metal pollution and treatment. Since the twenty-first century, the concentration of heavy metals in wetland soil has often exceeded the threshold of soil protection law, and the wetland is often used as hay production land and cattle farm. Concerns about the possible transfer of heavy metals into the food web have been raised. Early scholars studied the absorption and transport of heavy metals by soil and vegetation in wetlands. Asa et al. found that the heavy metals in the stems and leaves of Potamogeton natans were directly absorbed from water, and there was no apparent transfer from root to bud (Fritioff and Greger 2006). Liu et al. studied the absorption of Cd, Pb, and Zn by 19 wetland plants, and the results showed that the removal rate of heavy metals was more than 90% (Liu et al. 2007). The above research indicated that plant extraction would be a better remediation technology for heavy metal pollution. However, most plants experience low bio effectiveness or limited metal transfer to above-ground parts during extraction. Researchers began to study the use of chemicals to assist plant extraction of metals, synthetic chelating agents, and low molecular weight organic acids often used to improve the bioavailability of metals, which are the focus of research. Studies have confirmed that citric acid can increase the absorption of Cd, Mn, Pb, and Hg by plants more effectively than other organic acids with low molecular weight. At the same time, EDTA can improve the accumulation of metals in plant buds. However, slow degradation and longer duration in the soil will increase the environmental leaching risk and toxicity, so it is not suitable for practical use (Amir et al. 2020).

In recent years, the research focused on heavy metals in the wetland is the source analysis and risk assessment of heavy metals. Sabry et al. assessed heavy metal pollution in the sediments of the Brulus coastal lagoon by using pollution factors, pollution load index, and other methods and showed that the sediments were seriously polluted by metals such As and Sb (Shaheen et al. 2019). Heavy metals in sediments are likely to accumulate gradually through the food chain in animals and humans. Yu et al. have shown that sandpipers in wetlands, small birds that forage from sediments, are susceptible to copper and lead (Man et al. 2021). Wang et al. found that the pollution levels of As, Cd, and Zn in the riparian wetlands in Yunnan were relatively high. Contamination levels of other elements are relatively low, probably due to the different bioavailability of each metal (Wang et al. 2018). In the study of Wang et al. (2019), Diffusive Gradients in Thin-films (DGT), High-resolution Peeper (HR-Peeper), and traditional methods were combined to evaluate heavy metals in wetland sediments in Taihu Basin. DGT technology can reflect the dynamic supply of metals in the solid phase and accurately reflect the mobility and bioavailability of metals in sediments. The methods used to assess heavy metal pollution levels in various studies have been used for decades, which may not accurately assess the risks existing in the current wetland environment. Therefore, the risk assessment of heavy metals in wetlands should consider factors such as wetland type and metal bioavailability. Meanwhile, a database of heavy metal concentrations should be established according to different wetland ecosystems to facilitate the study of heavy metal bioavailability and transport in different ecosystems.

Control of nutrients and organic pollutants. The above co-occurrence and emergent analysis results show that constructed wetlands are widely used to manage nutrients and organic pollutants. Since 2005, researchers have begun to pay attention to the ability of constructed wetlands to remove nutrients and improve the ability of constructed wetlands to remove pollutants by testing different types of constructed wetlands, vegetation, hydraulic loads, and medium materials. If the electrochemical method is applied to the constructed wetland, the nitrate removal rate can be increased to more than 80%, and the phosphorus removal rate can reach more than 90% (Gao et al. 2017). The nitrogen removal rate reached 98.3% by adding biochar in tidal flow mode (Li et al. 2019). The removal rate of high chlorine compounds in the horizontal flow constructed wetland is up to 99%. However, the removal rate of organic pollutants containing less chlorine or no chlorine is low. The interaction between pollutants will seriously affect their removal effect, which is a crucial point worth attending to in the future (Ávila et al. 2014). In recent years, wetland organic pollutants have been the focus of attention, especially emerging organic pollutants. Early studies on organic pollutants in wetlands focused on the distribution characteristics, source analysis, and potential harm degree of PAHs in sediments and fish. Several areas have shown that the hazards of organic pollutants to wetland organisms and humans are within acceptable limits. In the long run, managing organic pollutants in the wetland is still worthy of attention. Current research focuses on the remediation of PAHs contaminated soils through physical, chemical, and biological techniques. Physical and chemical remediation, although fast, is prone to secondary contamination and soil deactivation and is expensive. Biodegradation is considered a simple, environmentally friendly, and low-cost method. However, more studies need to be done on the biodegradation of organic pollutants. Moreover, biodegradation is mainly at the stage of basin experiments, and the removal effect is common. Nano and transgenic technologies have been proven effective in removing heavy metals and organic pollutants, but few studies have applied them in wetland management. In the future, biotechnology can be combined with nanotechnology and transgenic technology to be used in the research of removing organic pollutants.

8.5 Conclusion

In this paper, we analyzed the temporal evolution, cooperation, and research hotspots of wetland environmental science literature in WOS from 1977 to 2021 using the bibliometric analysis software CiteSpace, and revealed the research hotspots and future research directions in the field of wetland environmental science. The conclusions are as follows:

(1) The number of publications in wetland environmental science has shown a rapid growth trend after 2005 and is expected to continue to grow in the future; the United States and China have contributed much more than other countries in this field. China's research started late but has been multiplying in the past five years.
(2) According to the co-analysis of authors, Joan García is the author with the highest number of publications in the field of wetland environmental science, followed by Cui Baoshan, Bai Junhong, Hans Brix, and Zhang Jian. The high output was obtained in the research direction of artificial wetlands for wastewater removal.
(3) In the analysis of keywords, the emergent analysis found that the wetland environmental science research hotspot showed three stages of change over time. The first stage was the quantitative analysis of wetland greenhouse gases. The second stage is the interaction of heavy metal pollution, climate change, and wetland ecosystem. The third stage is the impact of human or natural activities

on wetlands and the degree of wetland degradation and restoration measures. Cluster analysis shows that pattern, atmospheric methane, and mercury are the most extensive study scale clusters. Based on the authors' cooperation analysis and the emergence analysis, the cluster labels can be divided into two categories: the relationship between wetlands and climate change and wetland pollution and control. In the future, detailed classification of wetlands, exploration of the determinants of various wetland ecosystems, the combination of bioremediation technology and nanotechnology or transgenic technology, and wetland restoration measures may be the key to scientific research on the wetland environment.

In terms of data sources, retrieval methods, research methods, and analysis results, this paper has a particular reference significance for the research trends in wetland environmental science. Bibliometric analysis can objectively and quantitatively reflect the research trends in the field of wetland environmental science at the macro level, but there are certain limitations in the research methods. In the follow-up research, the software function should be further improved, or other methods should be used.

References

Adhikari K, Owens PR, Libohova Z, Miller DM, Wills SA, Nemecek J (2019) Assessing soil organic carbon stock of Wisconsin, USA and its fate under future land use and climate change. Sci Total Environ 667:833–845

Amir W, Farid M, Ishaq HK, Farid S, Zubair M, Alharby HF, Bamagoos AA, Rizwan M, Raza N, Rehman K, Ali HS (2020) Accumulation potential and tolerance response of Typha latifolia L. under citric acid assisted phytoextraction of lead and mercury. Chemosphere 257:127247

Ávila C, Matamoros V, Reyes-Contreras C, Piña B, Casado M, Mita L, Rivetti C, Barata C, García J, Bayona JM (2014) Attenuation of emerging organic contaminants in a hybrid constructed wetland system under different hydraulic loading rates and their associated toxicological effects in wastewater. Sci Total Environ 470:1272–1280

Bai JH, Xiao R, Cui BS, Zhang KJ, Wang QG, Liu XH, Gao HF, Huang LB (2011) Assessment of heavy metal pollution in wetland soils from the young and old reclaimed regions in the Pearl River Estuary South China. Environ Pollut 159(3):817–824

Baldwin AH, Jensen K, Schönfeldt M (2014) Warming increases plant biomass and reduces diversity across continents, latitudes, and species migration scenarios in experimental wetland communities. Glob Change Biol 20(3):835–850

Bohn TJ, Lettenmaier DP, Sathulur K, Bowling LC, Podest E, McDonald KC, Friborg T (2007) Methane emissions from western Siberian wetlands: heterogeneity and sensitivity to climate change. Environ Res Lett 2(4):045015

Bonanno G, Giudice RL (2010) Heavy metal bioaccumulation by the organs of Phragmites australis (common reed) and their potential use as contamination indicators. Ecol Ind 10(3):639–645

Chen D, Liu XY, Bian RJ, Cheng K, Zhang XH, Zheng JF, Josepha S, Crowley D, Pan GX, Li LQ (2018) Effects of biochar on availability and plant uptake of heavy metals—A meta-analysis. J Environ Manag 222:76–85

Fritioff Å, Greger M (2006) Uptake and distribution of Zn, Cu, Cd, and Pb in an aquatic plant Potamogeton natans. Chemosphere 63(2):220–227

Gao Y, Xie YW, Zhang Q, Wang AL, Yu YX, Yang LY (2017) Intensified nitrate and phosphorus removal in an electrolysis-integrated horizontal subsurface-flow constructed wetland. Water Res 108:39–45

He CY, Liu ZF, Tian J, Ma Q (2014) Urban expansion dynamics and natural habitat loss in China: a multiscale landscape perspective. Glob Change Biol 20(9):2886–2902

Hu SJ, Niu ZG, Chen YF, Li LF, Zhang HY (2017) Global wetlands: potential distribution, wetland loss, and status. Sci Total Environ 586:319–327

Kuhfuss L, Rey-Valette H, Sourisseau E, Heurtefeux H, Rufray X (2016) Evaluating the impacts of sea level rise on coastal wetlands in Languedoc-Roussillon, France. Environ Sci Policy 59:26–34

Li J, Hu Z, Li F, Fan J, Zhang J, Li F, Hu H (2019) Effect of oxygen supply strategy on nitrogen removal of biochar-based vertical subsurface flow constructed wetland: intermittent aeration and tidal flow. Chemosphere 223:366–374

Li CM, Wang HC, Liao XL, Xiao R, Liu KH, Bai JH, Li B, He Q (2022) Heavy metal pollution in coastal wetlands: a systematic review of studies globally over the past three decades. J Hazard Mater 424:127312

Liang J, Yang ZX, Tang L, Zeng GM, Yu M, Li XD, Wu HP, Qian YY, Li XM, Luo Y (2017) Changes in heavy metal mobility and availability from contaminated wetland soil remediated with combined biochar-compost. Chemosphere 181:281–288

Liu J, Dong Y, Xu H, Wang D, Xu J (2007) Accumulation of Cd, Pb and Zn by 19 wetland plant species in constructed wetland. J Hazard Mater 147(3):947–953

Man YB, Chow KL, Zhang F, Lei KM, Leung AOW, Mo WY, Wong MH (2021) Protecting water birds of wetlands: Using toxicological tests and ecological risk assessment, based on metal/loid (s) of water, sediment and biota samples. Sci Total Environ 778:146317

Mohamed YA, Van den Hurk BJJM, Savenije HHG, Bastiaanssen WG M (2005) Impact of the Sudd wetland on the Nile hydroclimatology. Water Resour Res 41(8)

Pickett-Heaps CA, Jacob DJ, Wecht KJ, Kort EA, Wofsy SC, Diskin GS, Worthy DEJ, Kaplan JO, Bey I, Drevet J (2011) Magnitude and seasonality of wetland methane emissions from the Hudson Bay Lowlands (Canada). Atmos Chem Phys 11(8):3773–3779

Ravikumar Y, Yun J, Zhang G, Zabed HM, Qi X (2022) A review on constructed wetlands-based removal of pharmaceutical contaminants derived from non-point source pollution. Environ Technol Innov 26:102504

Shaheen SM, Abdelrazek MA, Elthoth M, Moghanm FS, Mohamed R, Hamza A, El-Habashi N, Wang JX, Rinklebe J (2019) Potentially toxic elements in saltmarsh sediments and common reed (Phragmites australis) of Burullus coastal lagoon at north Nile Delta, Egypt: a survey and risk assessment. Sci Total Environ 649:1237–1249

Song C, Xu X, Tian H, Wang Y (2009) Ecosystem–atmosphere exchange of CH4 and N2O and ecosystem respiration in wetlands in the Sanjiang plain, northeastern China. Global Change Biol 15(3):692–705

Staszak LA, Armitage AR (2013) Evaluating salt marsh restoration success with an index of ecosystem integrity. J Coastal Res 29(2):410–418

Turetsky MR, Kotowska A, Bubier J, Dise NB, Crill P, Hornibrook ER, Minkkinen K, Moore TR, Myers-Smith IH, Nykänen H, Olefeldt D, Rinne J, Saarnio S, Shurpali N, Tuittila E, Waddington JM, White JR, Wickland KP, Wilmking M (2014) A synthesis of methane emissions from 71 northern, temperate, and subtropical wetlands. Global Change Biol 20(7):2183–2197

Wang Z, Hou L, Liu Y, Wang Y, Ma LQ (2018) Metal contamination in a riparian wetland: distribution, fractionation and plant uptake. Chemosphere 200:587–593

Wang W, Wang S, Chen J, Jiang X, Zheng B (2019) Combined use of diffusive gradients in thin film, high-resolution dialysis technique and traditional methods to assess pollution and bioavailability of sediment metals of lake wetlands in Taihu Lake Basin. Sci Total Environ 671:28–40

Wang WD, Yang T, Guan WB, Peng WX, Wu P, Zhong B, Zhou CD, Chen QH, Zhang RB, Xu KW, Yin CQ (2021) Ecological wetland paradigm drives water source improvement in the stream network of Yangtze River Delta. J Environ Sci 110(12):55–72

Worthy DE, Levin I, Hopper F, Ernst MK, Trivett NB (2000) Evidence for a link between climate and northern wetland methane emissions. J Geophys Res: Atmos 105(D3):4031–4038

Wu TL, Wang YJ, Chen HM (2021) Research progress and hotspots of environmental soil science between 2016–2020 based on bibliometrics analysis. J Agro-Environ Sci 40(1):1–15

Yi X, Tran NH, Yin T, He Y, Gin KYH (2017) Removal of selected PPCPs, EDCs, and antibiotic resistance genes in landfill leachate by a full-scale constructed wetlands system. Water Res 121:46–60

Zhao XY, Qin HM (2022) Progress, hotspots and prospects for Zooplankton research in China over the past two decades (1999–2019). J Hydroecol 43(4):39–44

Zhu W, Xiang XQ, Hou LP, Wang BS, Tang LN (2018) Knowledge mapping analysis of ecological risk research based on Citespace. Acta Ecol Sin 38(12):4504–4515

Chapter 9
Calculation for Resistance Loss of Fine Dredged Slurry in Long-Distance Straight Pipeline

Rui Wang and Guoliang Yu

Abstract In order to improve the protection of water resources and the quality of the environment and ecology, a huge amount of dredging business was carried out every year, such as the excavation and maintenance dredging of harbors and waterways, the construction of coastal industrial parks, artificial islands, tourist ecological resorts, and the dredging to maintain the capacity and safety of reservoirs. In this study, the calculation for the resistance loss of the fine dredged slurry in long straight pipeline was investigated based on the calculation method of the flow resistance in open channels. By introducing the drag coefficient, the calculation model for the resistance loss of dredged slurry in long straight pipeline was development, which should be divided into high energy area and low energy area. The commonly used calculation models and the measured results in previous research were used to verify the accuracy of the calculation model proposed in this paper. The verification results show that the calculation model for the resistance loss of the dredged slurry in long straight pipeline proposed in this paper has a reliable accuracy, which can reflect the non-monotonic change of the flow resistance loss in the pipeline, and has good practicability.

Keywords Resistance loss · Fine dredged slurry · Long-distance straight pipeline · Drag coefficient · Calculation model

9.1 Introduction

Long-distance slurry transport in pipeline is one of the main ways of the solid hydraulic pipeline transport in dredging industry (Li 2011). More than 90% of the large dredging projects are excavated by cutter suction dredgers or rake suction dredgers, and then the excavated slurry is transported to destination by a pipeline (Goes et al. 2018; Nanda 2019). In the process of dredging, the transport velocity of

R. Wang · G. Yu (✉)
School of Naval Architecture, Ocean and Civil Engineering, Shanghai Jiao Tong University, Shanghai 200240, China
e-mail: yugl@sjtu.edu.cn

H. Xu (ed.), *Proceedings of the 5th International Symposium on Water Resource and Environmental Management*, Environmental Science and Engineering,
https://doi.org/10.1007/978-3-031-31289-2_9

slurry in the pipeline needs to reach a certain value to keep the slurry suspended, otherwise the slurry will settle, which results in the increase of slurry transport resistance (Li et al. 2022), and reduced the transport efficiency. The design and the operation of hydraulic transmission system also need to estimate the slurry resistance loss in the pipeline reliably (Sun et al. 2021). The slurry resistance loss determines the type of dredging equipment (Tian et al. 2021) and the size of the pipeline (Wang et al. 2016), and is related to the cost and the efficiency of slurry pipeline transportation (Guo et al. 2020). Thus, the calculation of slurry resistance loss in dredged pipeline is a basic problem in dredging industry.

The internal mechanism of the slurry resistance loss in pipeline transportation lies in the energy dissipation along the pipeline, which is mainly manifested as the pressure reduction (Matousek 2002, 2009) or the hydraulic loss along the pipeline (Ma et al. 2015; Gao et al. 2016). Its composition and characteristics from the macroscopic and microscopic perspective include the friction resistance loss between slurry and pipe wall (Yu et al. 2012), the settlement resistance loss and the collision resistance loss of sediment particle (Zhao et al. 2019). Because fine slurry has a certain viscosity, its concentration and physicochemical properties may also change in the process of dredging pipeline transportation (Wang and Yu 2019), and the flow pattern changes have various forms (Yang et al. 2014), so far it is impossible to conduct a complete unified analysis theoretically. Therefore, many empirical formulas are mostly adopted at home and abroad for mathematical calculation for the slurry transport resistance loss in pipeline transportation (Durand 1953; Turian and Yuan 1977; Sun et al. 2004). However, these formulas are of low accuracy when they are applied to the calculation for the transport resistance of high-concentration fine slurry in pipeline transportation.

In this paper, the calculation model for the resistance loss of the fine dredged slurry in long straight pipe-line was developed referring to the calculation method of the flow resistance in open channel firstly. The influence of bedforms were considered. Finally, the experimental results and the field test result were collected to verify the accuracy of the calculation model proposed in this paper. This research provided a new idea and method for the design of slurry pipeline transportation and the study of slurry resistance loss in slurry pipeline transportation.

9.2 Calculation Model Derivation for Resistance Loss

9.2.1 Different Bed Energy Regions

The calculation model derivation for the resistance loss of the fine dredged slurry in long straight pipeline should consider the different energy regions of bed, because the bed form is an important factor affecting the transport resistance of the fine dredged slurry in long straight pipeline. When the fine dredged slurry is transported in a long straight pipeline, there will be some sand waves with different sizes and forms near bed. Whether there is sand waves in pipeline or not, the formation mechanism of the transport resistance of fine dredged slurry are different, which is reflected in the low energy region (the grain and dune stage) and the high energy region (the dynamic flat stage and sand wave stage). As a result, the calculation methods for the resistance loss of fine dredged slurry in different energy region in long straight pipeline are different. The influence of bed forms on the transport resistance of the dredged slurry in long straight pipeline is to change friction resistance loss and collision resistance (Li et al. 2018), and causes the difference in the concentration distribution and the movement of the fine dredged slurry on the section of the pipeline, which in turn will cause the transformation of the resistance coefficient in the calculation model for the resistance loss of the fine dredged slurry in long straight pipeline (Yu and Lim 2003). While the commonly used calculation models, such as the Durand model and Turian and Yuan model, don't consider this point, which often have a large error in the practical engineering applications. It is necessary to first determine the drag coefficient in different bed forms firstly.

9.2.2 Resistance Coefficient Calculation Method

Referring to the calculation method for the flow resistance in different energy states in open channel proposed by Yu and Lim (2003), the average slurry transport velocities under the conditions of the impact bed with sand waves present and the rigid bed with no sand waves present in the long straight pipe were defined as U and U_r respectively, and the resistance coefficient λ was the ratio of U and U_r, which was derived as follows.

$$\lambda = \frac{U}{\xi U_r} \tag{9.1}$$

ξ in the above equation represented the influence of sediment characteristics and bed characteristics on the transport resistance of fine dredged slurry in long straight pipeline, and the function expression was derived as follows.

$$\xi = \begin{cases} 1 & R_b/d_{50} \geq 300\chi \\ \dfrac{300\chi}{R_b/d_{50}} & R_b/d_{50} \leq 300\chi \end{cases} \tag{9.2}$$

where R_b was the hydraulic radius of pipeline, m; d_{50} was the sediment particle size, m; χ was the dimensionless form of bed shear stress, which was derived as follows.

$$\chi = \ln \theta / \theta_{cr} \tag{9.3}$$

where θ and θ_{cr} were the actual Shields number and the critical Shields number when the fine dredged slurry was transported in a long straight pipeline.

$$\theta = \gamma_m DJ/4\theta_{cr} \quad (9.4)$$

$$\theta_{cr} = 0.056 - 0.33e^{-0.0115d_{gr}} + 0.1e^{-0.25d_{gr}} + e^{-2d_{gr}} \quad (9.5)$$

where γ_m was the bulk density of fine dredged slurry, N/m^3; D was the pipe diameter, m; J was the hydraulic gradient; d_{gr} was the dimensionless form of sediment particle size, which was derived as follows.

$$d_{gr} = d_{50}[g(\gamma_s - \gamma)/\gamma\nu^2]^{1/3} \quad (9.6)$$

where ν is the kinematic viscosity, m^2/s.

According to the study of Yu and Lim (2003), the resistance coefficient λ is correlated with the dimensionless form of bed shear stress χ. $\theta/\theta_{cr} \le 1$ represents the flow intensity is lower than the incipient condition of bed sediment, and the bed form is flat. At this time, $\lambda = 1$, but in the normal slurry pipeline transportation process, the flow intensity is always much greater than the incipient condition of bed sediment, which does not belong to the scope of this study. For the low energy region, with the increase of θ/θ_{cr}, the resistance coefficient λ decreases with the formation of ripples and dunes. When $1 \le \theta/\theta_{cr} \le 7.5$, the relationships between the resistance coefficient λ and the dimensionless form of bed shear stress χ in the low energy region and the high energy region were consistent. While $7.5 \le \theta/\theta_{cr} \le 250$, the relationships between the resistance coefficient λ and the dimensionless form of bed shear stress χ in the low energy region and the high energy region were very different. For $1 \le \theta/\theta_{cr} \le 250$, the relationship between the resistance coefficient λ_L and the dimensionless form of bed shear stress χ in the low energy region was developed as follows (Fig. 9.1).

$$\lambda_L = -0.0044\chi^3 + 0.0661\chi^2 - 0.352\chi + 1 \quad (9.7)$$

For the high energy region, the relationship between the resistance coefficient λ_H and the dimensionless form of bed shear stress χ was divided into two parts according to the value of θ/θ_{cr}. When $1 \le \theta/\theta_{cr} \le 7.5$,

$$\lambda_H = -0.0044\chi^3 + 0.0661\chi^2 - 0.352\chi + 1 \quad (9.8)$$

And when $7.5 \le \theta/\theta_{cr} \le 250$,

$$\lambda_H = 0.0337\chi - 0.469\chi^2 + 1.916\chi - 1.644 \quad (9.9)$$

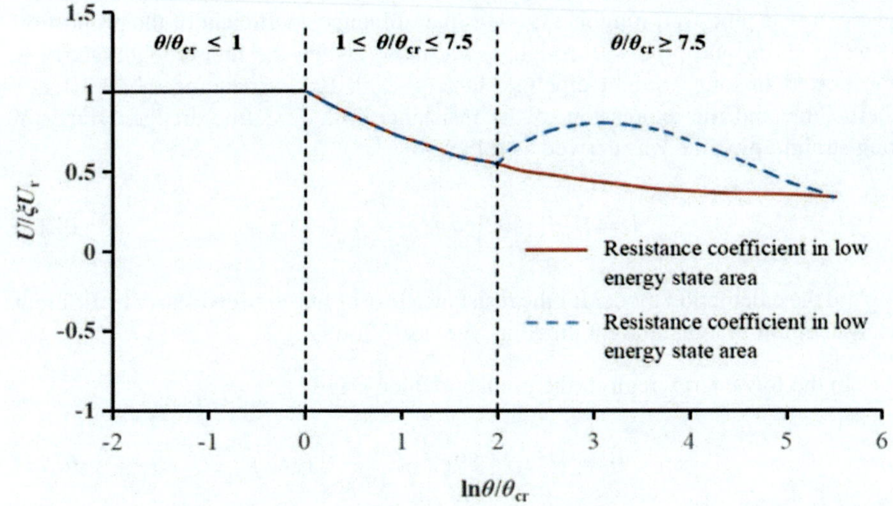

Fig. 9.1 Relationship between resistance coefficient and dimensionless form of bed shear stress

9.2.3 *Resistance Loss Calculation Model*

The hydraulic slope J was used to represent the resistance loss of the fine dredged slurry in long straight pipeline, because the resistance loss of the fine dredged slurry in long straight pipeline was mainly manifested as the pressure reduction or the hydraulic loss along the way. When there were sand waves in pipeline, the average sectional velocity of slurry in the long straight pipeline was calculated by Manning formula, as follows.

$$U = \frac{1}{n} R_b^{\frac{2}{3}} J^{\frac{1}{2}} \tag{9.10}$$

where R_b was the hydraulic radius of pipeline, $R_b = 1/4D$, n was the Manning coefficient, $n = (d_{50})^{1/6}/6.7g^{1/2}$, and g was the acceleration of gravity, m/s^2. According to the study of Yu and Lim (2003), the Manning coefficient n could be substituted into the Eq. (9.1) to obtain the average velocity of slurry in the pipeline under the condition of the bed with non-erosive materials or the bed with loose material, when the flow intensity in pipeline was small, and the bed form was flat. Therefore, the formula for the average sectional velocity of slurry in the long straight pipeline under the condition of the bed with no sand waves was derived as follows.

$$U_r = 6.67\sqrt{gd_{50}}\left(\frac{D}{4\eta\Delta}\right)^{\frac{2}{3}} J^{\frac{1}{2}} \tag{9.11}$$

where Δ was pipe wall roughness; η was the influence coefficient of the proportion between slurry and pipe wall roughness. Usually, when the fine dredged slurry is transported in long straight pipeline, $D/\Delta \geq 500$. By introducing the resistance coefficient, and the expression of the resistance loss J of fine dredged slurry in long straight pipeline was derived as follows.

$$J = U^2/[44.89gd_{50}(\frac{D}{4\eta\Delta})^{\frac{4}{3}}(\xi\lambda)^2] \tag{9.12}$$

And the calculation model for the resistance loss of fine dredged slurry in different energy region in long straight pipeline were as follows.

(1) In the low energy region (the grain and dune stage):

$$J_L = U^2/[44.89gd_{50}(\frac{D}{4\eta\Delta})^{\frac{4}{3}}(\xi\lambda_L)^2] \tag{9.13}$$

(2) In the high energy state area (the dynamic flat stage and sand wave stage):

$$J_H = U^2/[44.89gd_{50}(\frac{D}{4\eta\Delta})^{\frac{4}{3}}(\xi\lambda_H)^2] \tag{9.14}$$

9.3 Calculation Model Test and Discussion

9.3.1 Verification by Measured Data

The field measured data and the experimental data collected by Gong et al. (2021) about the transport velocity of the fine dredged slurry in long straight pipeline were used to verified the calculation model proposed in this paper. The sediment parameters and the hydraulic conditions are shown in Table 9.1. According to the data in Table 9.1, R_b/d_{50}, χ and ξ of the dredged slurry in this pipeline transportation were calculated. The calculation results shown that R_b/d_{50} was always smaller than 300χ, such as all the data of the sediment with the particle sizes 0.07 mm, 0.15 mm, 0.11 mm and part of the data of the sediment with the particle size 0.056 mm. In this case, the value of coefficient ξ is not 1, which should be obtained according to formula (9.2). The condition that $R_b/d_{50} \geq 300\chi$ mostly occurred when the sediment particle size is less than or equal to 0.06 mm, such as all the data of the sediment with the particle size 0.027 mm and part of the data of the sediment with the particle size 0.056 mm. In this case, the value of coefficient ξ was 1. The reason for $R_b/d_{50} \geq 300\chi$ was that the influence of the cohesive force between sediment particles on the resistance loss of slurry pipeline transportation gradually increased, when the sediment particle size was less than 0.06 mm.

Table 9.1 Field measurement data of slurry transportation

Pipe diameter (mm)	Sediment density (g/cm^3)	Sediment particle size (mm)	Volume concentration (%)	Flow intensity	Incipient condition	R_b/d_{50}
268	1.33	0.07	13	10.51	0.034	957.14
268	1.33	0.07	13	7.80	0.034	957.14
268	1.33	0.07	13	4.91	0.034	957.14
321	1.44	0.15	30.3	8.53	0.086	535
321	1.44	0.15	30.3	5.65	0.086	535
321	1.44	0.15	30.3	17.95	0.086	535
321	1.44	0.15	30.3	5.10	0.086	535
248	1.35	0.056	19.8	40.99	0.291	1107.14
248	1.35	0.056	19.8	7.91	0.291	1107.14
248	1.35	0.056	19.8	8.78	0.291	1107.14
248	1.35	0.056	19.8	6.96	0.291	1107.14
255	1.01	0.027	6.3	13.21	0.9	2361.11
255	1.01	0.027	6.3	15.99	0.9	2361.11
255	1.01	0.027	6.3	7.32	0.9	2361.11
187	1.61	0.11	14.5	11.34	0.104	425
187	1.61	0.11	14.5	8.67	0.104	425
187	1.61	0.11	14.5	17.6	0.104	425
187	1.61	0.11	14.5	15.3	0.104	425

In this paper, the calculated transport velocity of the fine dredged slurry in long straight pipeline was adopted to be compared and verified with the measured transport velocity, which was a more intuitive comparison. The resistance loss wasn't adopted as the index because the measured value of the resistance loss of the fine dredged slurry in long straight pipeline was too small, which couldn't well reflect the comparison effect with the calculated results. The calculated transport velocity of the fine dredged slurry in long straight pipeline was obtained by putting the measured hydraulic slope into the calculation models. In addition, the bed form is rarely in the stage of dynamic flatness and sand waves during the fine dredged slurry transporting in long straight pipeline. Thus, the calculation model in the low energy region proposed in this paper was used to obtain the transport velocity of fine dredged slurry in long straight pipeline.

Figure 9.2 describes the comparison between the field measured results and the calculation results. It could be seen from Fig. 9.2 that the calculated results were close to the measured data, which were basically within the range of 10% error, and evenly distributed within the range of the error line. It was proved that the proposed calculation model had a good accuracy.

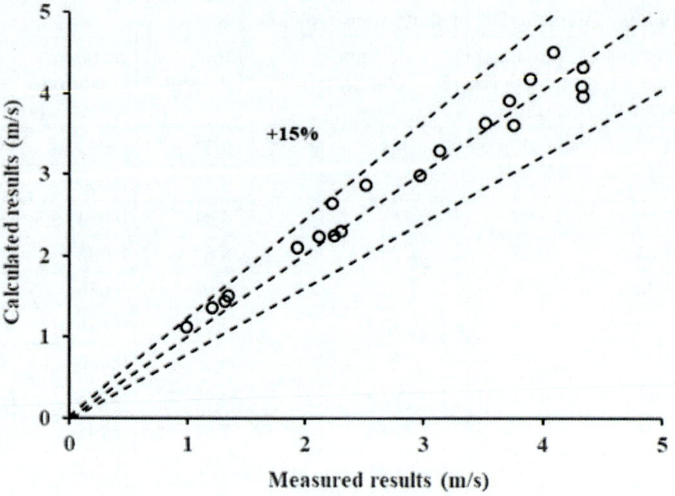

Fig. 9.2 Comparison between the field measured results and the calculation results

9.3.2 Comparison with Commonly Used Calculation Models

The calculation model proposed in this paper was compared and analyzed with the commonly used calculation models based on the measured data of the pipeline transportation in Jianshan iron concentrate of Shanghai and Nanfen tailings of Liaoning province in China. The following four commonly used calculation models adopted, which were listed as follows.

(1) Durand model

Durand model proposed by Durand (1953) expresses the resistance loss as the sum of the resistance loss in clear water and the additional resistance loss by introducing the concept of critical deposition velocity required to the maximum effective operating condition, which is still used in the industrial field.

$$J = J_W + K_D C_V J_W [\frac{V_m^2 \sqrt{C_D}}{gD(\delta - 1)}]^{-\frac{3}{2}} \tag{9.15}$$

where K_D is the constant coefficient, $K_D = 82$; C_v is the volume concentration of slurry, %; V_m is the average flow velocity of the solid–liquid mixture (slurry), m/s; C_D is the drag coefficient; δ is the ratio of the solid (slurry) density to the liquid (water) density. J_W is the hydraulic slope of clear water, which is derives as follows.

$$J_W = \varepsilon V_M^2 / 2gD \tag{9.16}$$

$$\varepsilon = 124.5N^2/D^{1/3} \tag{9.17}$$

where N is the roughness coefficient.

(2) Turian and Yuan model

Turian and Yuan model proposed by Turian and Yuan (1977) believes that the hydraulic slope of the slurry in pipeline transportation is equal to the sum of the hydraulic slope in clean water and the additional hydraulic slope generated by the intervention of solid particles, which is derives as follows.

$$J = J_W + 0.551C_V^{0.869}C_D^{-0.168}[\frac{V_m^2}{gd}(\delta - 1)]^{-0.694}J_W^{1.2} \tag{9.18}$$

where $C_D = 0.4 \sim 0.43$, when $1000 < R_{ed} < 2 \times 10^5$, and R_{ed} os the particle Reynolds number.

(3) Calculation model proposed by Wenguang Chen

According to the calculation model proposed by Chen (1994), the total resistance loss of the slurry in pipeline transportation is the superposition of the friction, the particle sedimentation and the particle collision, which is shown as follows.

$$J = \frac{\lambda V_m^2}{gd} + C_V(\delta - 1)\frac{\omega}{V_m} + \frac{d_{50}V_m}{D^2} \cdot \frac{C_V}{(1 - C_V/C_{Vm})^{2.5C_V}} \tag{9.19}$$

where C_{Vm} is the limiting volume concentration of slurry, %; ω is the settling velocity of sediment particles, m/s. The following formula is adopted to obtain the settling velocity of sediment particles, which is applicable to the calculation for the settling velocity of sediment particles in laminar, turbulent and transition zones (Ni et al. 2004), and as follows.

$$\omega = \sqrt{(13.95\frac{\nu}{d_{50}})^2 + 1.09\frac{\gamma_s - \gamma}{\gamma}gd_{50}} - 13.95\frac{\nu}{d_{50}} \tag{9.20}$$

where γ_s and γ are the bulk density of sediment particles and clear water respectively, N/m^3.

(4) Calculation model proposed by Xiangjun Fei

The calculation model proposed by Fei (1986) decomposes the transport resistance of slurry in pipeline transportation into the carrier resistance and the bed resistance, which is derived as follows.

$$J = \frac{\lambda V_m^2\gamma_m}{2gD\gamma} + 11\mu_s C_V(\frac{\gamma_s - \gamma}{\gamma})\frac{\omega}{V_m} \tag{9.21}$$

Table 9.2 Hydraulic conditions of pipe transportation in Jianshan iron concentrate

Pipe diameter (mm)	Sediment density (g/cm^3)	Sediment particle size (mm)	Volume concentration (%)	Flow intensity	Incipient condition
255	4.76	0.027	17	11.51	0.271
255	4.76	0.027	17	6.80	0.271
255	4.76	0.027	17	3.91	0.271
255	4.76	0.027	11.8	16.37	0.271
255	4.76	0.027	11.8	9.53	0.271
255	4.76	0.027	11.8	4.65	0.271
100	4.76	0.027	17.1	18.95	0.271
100	4.76	0.027	17.1	3.10	0.271
100	4.76	0.027	33.2	48.99	0.271
100	4.76	0.027	33.2	3.91	0.271

where μ_s is the friction coefficient of bed surface, and the value of μ_s is related to the sediment particle properties, which is usually adopted 0.37 ~ 0.75; γ_m is the bulk density of slurry, N/m^3.

The hydraulic conditions of pipeline transportation in Jianshan iron concentrate of Shanghai and Nanfen tailings of Liaoning province in China. are shown in Tables 9.2 and 9.3. Based on the data in Tables 9.2 and 9.3, R_b/d_{50}, χ and ξ of the pipeline transportation in Jianshan iron concentrate and Nanfen tailings were calculated respectively. Different from the sediment transport in open channel flow, the values of R_b/d_{50} in pipeline transportation were mostly less than 300 χ. For example, the measured data in lines 7–10 of Table 9.2 and the measured data in lines 1–16 of Table 9.2. In this case, the value of coefficient ξ is not 1 and should be obtained according to formula (9.2). There would also be the cases in slurry pipeline transportation that R_b/d_{50} was greater than 300χ, in which case the value of coefficient ξ is 1.

The hydraulic slope (transport resistance) was estimated by using the calculation model proposed in this paper and the commonly used models, and then the slurry transport velocity of the pipeline transportation in Jianshan iron concentrate of Shanghai and Nanfen tailings of Liaoning province in China was obtained respectively. The calculated results were verified with the field measured results. The calculated results and the field measured results are shown in Tables 9.4 and 9.5.

It could be seen from Tables 9.4 and 9.5 that the calculation model proposed in this paper processed good calculational accuracy for the resistance loss of fine dredged slurry in long straight pipeline; Durand model and Turian and Yuan model had good calculational accuracy when R_b/d_{50} was less than 300 χ ($\xi = 1$), but the calculation accuracy was worse than that of the calculation model proposed in this paper, which further verified that the calculated results of the Durand model was greater than the measured results in the region where the flow velocity was low, while the opposite was true in the region where the flow velocity was high, and also

Table 9.3 Hydraulic conditions of pipe transportation in Nanfen tailings

Pipe diameter (mm)	Sediment density (g/cm^3)	Sediment particle size (mm)	Volume concentration (%)	Flow intensity	Incipient condition
80	2.79	0.0646	8.2	16.91	0.125
80	2.79	0.0646	8.2	5.53	0.125
80	2.79	0.0646	13.3	19.06	0.125
80	2.79	0.0646	13.3	5.87	0.125
80	2.79	0.0646	19.3	16.49	0.125
80	2.79	0.0646	19.3	4.53	0.125
80	2.79	0.0646	26.4	17.59	0.125
80	2.79	0.0646	26.4	2.62	0.125
149	2.79	0.0646	8.2	17.84	0.125
149	2.79	0.0646	8.2	4.43	0.125
149	2.79	0.0646	13.3	17.62	0.125
149	2.79	0.0646	13.3	4.88	0.125
149	2.79	0.0646	19.3	17.43	0.125
149	2.79	0.0646	19.3	3.26	0.125
149	2.79	0.0646	26.4	14.80	0.125
149	2.79	0.0646	26.4	2.24	0.125

Table 9.4 Calculated results and the field measured results in Jianshan iron concentrate

Measured results (m/s)	Proposed calculation model (m/s)	Chen's model (m/s)	Durand model (m/s)	Turian and Yuan model (m/s)	Fei's model $\mu s = 0.37$ (m/s)
2.102	2.135	2.101	2.117	2.089	2.703
1.739	1.759	1.879	1.908	1.856	2.431
1.488	1.476	1.796	1.843	1.759	2.343
2.315	2.459	2.221	2.229	2.216	2.680
1.963	1.986	2.001	2.01533	1.991	2.424
1.589	1.554	1.828	1.856	1.808	2.233
2.00	2.236	1.902	1.86	1.859	2.385
1.00	1.182	1.281	1.303	1.257	1.672
3.00	3.376	2.703	2.587	2.596	3.874
1.00	1.262	1.287	1.284	1.199	1.862

Table 9.5 Calculated results and the field measured results in in Nanfen tailings

Measured results (m/s)	Proposed calculation model (m/s)	Chen's model (m/s)	Durand model (m/s)	Turian and Yuan model (m/s)	Fei's model $\mu s = 0.37$ (m/s)
2.18	2.152	2.052	1.877	1.884	2.016
1.25	1.343	1.300	1.193	1.198	1.297
2.36	2.268	2.219	2.015	2.025	2.25
1.3	1.375	1.362	1.233	1.241	1.398
2.17	2.128	2.085	1.873	1.887	2.184
1.19	1.244	1.319	1.171	1.179	1.399
2.13	2.190	2.077	1.831	1.849	2.235
0.93	1.026	1.200	1.014	1.018	1.31
2.5	2.444	2.339	2.215	2.223	2.383
1.39	1.368	1.480	1.413	1.417	1.552
2.57	2.430	2.418	2.28	2.293	2.552
1.4	1.419	1.491	1.409	1.415	1.621
2.52	2.419	2.390	2.239	2.256	2.616
1.19	1.224	1.384	1.286	1.287	1.587
2.2	2.252	2.144	1.982	2.003	2.433
0.99	1.082	1.314	1.177	1.162	1.586

specifically indicated that the calculational accuracy of the Durand model was in the acceptable range, but lower than the that of the calculation model proposed by Wenguang Chen; While the calculation results of the calculation model proposed by Xiangjun Fei were always larger than the measured results, and there still was a large error even if the friction coefficient μ_s was 0.37. In addition, the calculation accuracy of the calculation model proposed by Xiangjun Fei is lower when $R_b/d_{50} \leq 300\chi$ than that when $R_b/d_{50} \geq 300\chi$.

9.3.3 Discussion on Relationship Between Transport Velocity and Resistance Loss

The relationship curve between the slurry transport velocity and the slurry resistance loss was obtained by using the calculation model proposed in this paper. The value of the slurry resistance loss was taken as the as the X-axis, and the value of the slurry transport velocity was taken as the Y-axis, as shown in Fig. 9.3. It could be seen from Fig. 9.3 that the slurry transport velocity in pipeline always increased firstly, and then decreased with the increase of the resistance loss. There would be an inflection point in the relationship curve between the slurry transport velocity and the slurry resistance loss, The speed that the curve reached the inflection point

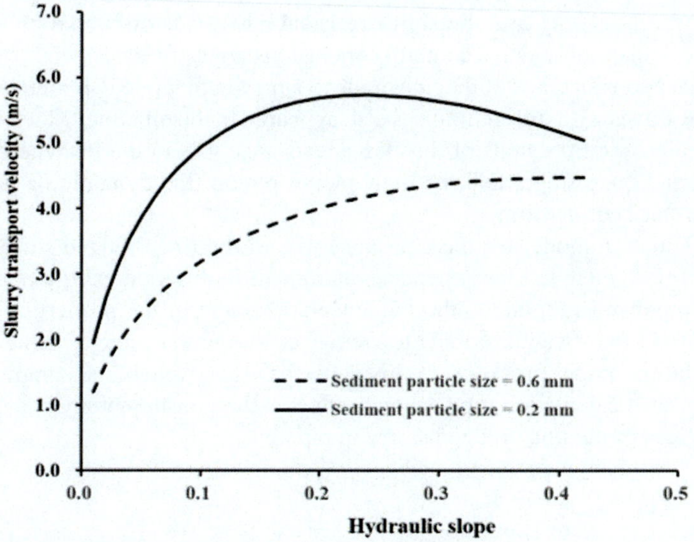

Fig. 9.3 Relationship curve between transport velocity and resistance loss

with the increase of the slurry resistance loss increased, and with the decrease of the sediment particle size. Before the inflection point appeared in the relationship curve, the increasing speed of the slurry transport velocity in pipeline decreased with the increase of sediment particle size.

This non-monotonic phenomenon between transport velocity and resistance loss has been proved by previous studies, such as Deng et al. (2005). In addition, the Durand model can also reflect the non-monotone change rule of the resistance loss in pipeline transportation of the medium coarse sediment in the low flow velocity zone. It is worth explaining that the reason for this phenomenon maybe that the bedform in the pipeline is transformed from large sand dunes to layers with the advance of slurry transportation, which results in the layer migration of bottom sediment, and thus reduces the transport resistance.

9.4 Summary and Conclusion

Referring to the calculation method of the flow resistance in open channel, the calculation model for the resistance loss of the fine dredged slurry in long straight pipeline was developed firstly. Then the field measured data of pipeline transportation was used to verify the calculational accuracy of the calculation model proposed in this paper. The commonly used calculation mode and the field measured data of pipeline transportation in Jianshan iron concentrate and Nanfen tailings were used to analyzed the superiority of the proposed calculation. Finally, the relationship between the slurry transport velocity and the slurry resistance loss was discussed by using the proposed calculation model. The main conclusions are as follows.

The transport resistance of the fine dredged slurry in pipeline was affected by bed form. Sand waves with different forms also appeared in the pipeline, which led to the different calculation methods of the slurry resistance loss in the low energy region (the grain and dune stage) and the high energy region (the dynamic flat stage and sand wave stage) of bedform.

The calculation model for the resistance loss of the fine dredged slurry in long straight pipeline included the resistance coefficient in different energy region, pipe diameter, pipe wall roughness, the influence coefficient of the proportion between slurry and pipe wall roughness and the average cross-section velocity in the pipeline. The calculation model proposed in this paper has more reliable accuracy than the commonly used calculation models, and reflected the non-monotony change of the resistance loss of the fine dredged slurry in pipeline.

References

Chen WG (1994) Experimental study on transport characteristics and vibration drag reduction of high concentration slurry. Doctoral dissertation of Central South University of Technology

Deng JX, Ni FS, Luo RM (2005) Two calculation models of sand transport resistance loss in pipeline. J Hohai Univ, Changzhou Branch 1:54–56

Durand R (1953) Basic relationships of the transportation of solids in pipes experimental research. In: 5th congress of the international association of hydraulic research. Minneapolic, Minnesota, pp 89–103

Fei XJ (1986) Prediction of hydraulic friction loss in solid pipeline. J Hydraul Eng 12:21–28

Gao J, Hao XD, Wang L, Wang P (2016) Resistance characteristics of coal slime in pipe flow at high pressure. Int J Chem Reactor Eng 14(1):299–307

Goes MG, Henriques MA, Cunha LVF, Gomes BPR, Guimaraes TAS (2018) Evaluation of internal corrosion in a Brazilian iron ore slurry pipeline based on the characterization of scales and tubercles. REM - Int Eng J 71(2):203–208

Gong XJ, Tao RL, Hao YC (2021) Critical velocity of sand transportation along long straight pipeline. Port Waterway Eng 8(535):13–23

Guo L, Zeng ZM, Huang XJ, Li MZ, Hao F, Li J, Rui XB (2020) Low-cost and high-efficiency method for detecting vertical bends of subsea pipelines. IEEE Access 8:33926–33933

Li YL (2011) A circulatory system and method of slurry delivery in long-distance pipeline transportation of iron ore. Proc SPIE—Int Soc Opt Eng 8205(1):82052S-82052S-3

Li MZ, He YP, Liu YD, Huang C (2018) Pressure-drop model of high-concentration graded particle transport in pipelines. Ocean Eng 163:630–640

Li MZ, He YP, Jiang RH, Zhang J, Zhang HS, Wei LH, Liu YD (2022) Analysis of minimum specific energy consumption and optimal transport concentration of slurry pipeline transport systems. Particuology 7:38–47

Ma S, Wang XD, Wu JD, Leng T, Ma J, Jiang D (2015) Study on resistance loss characteristics of iron concentrate pipeline transportation. In: The 27th Chinese control and decision conference. IEEE,15341233

Matousek V (2002) Pressure drops and flow patterns in sand-mixture pipes. Exp Thermal Fluid Sci 26:693–702

Matousek V (2009) Predictive model for frictional pressure drop in settling-slurry pipe with stationary deposit. Powder Technol 192:367–374

Nanda NK (2019) Slurry pipeline for transport of iron ore in India. Iron Steel Rev 63(6):86–89

Ni FS, Zhao LJ, Matousekv V (2004) Two-phase flow of high concentrated slurry in a pipeline. Dyn Res Progr 16(3):325–331

Sun DP, Wang EP, Yan J, Xu JG (2004) Research on resistance and transportation capacity of high concentration slurries transportation pipeline. J Hydraul Eng 35:93–99

Sun PY, Li Q, He HZ, Chen H (2021) Design and optimization investigation on hydraulic transmission and energy storage system for a floating-array-buoys wave energy converter. Energy Convers Manag 235(1):113998

Tian HL, Zhang XM, Jin G, Liang Y (2021) Study on the microstructure of anti-corrosion and wear resistant coating on the inner wall of equipment pipeline and the synthesis mechanism of ceramic coating. Mater Sci Forum 1026:163–168

Turian R, Yuan TF (1977) Flow of slurries in pipelines. AIChE J 23:232–243

Wang C, Deng GL, Zhou C, Yuan ZG (2016) Numerical simulation and experiment study on the resistance loss of glue pipeline transportation. In: International conference on electronic packaging technology. IEEE, 16357405

Wang R, Yu GL (2019) Incipient conditions of cohesive sediments in oscillatory flows. Ocean Coast Manag 181(1):104877

Yang WY, Yu GL, Tan SK, Wang HK (2014) Rheological properties of dense natural cohesive sediments subject to shear loadings. Int J Sedim Res 29(4):454–470

Yu GL, Lim SY (2003) Modified manning formula for flow in alluvial channels with sand-Beds. J Hydraul Res 41:597–608

Yu ZF, Tian XL, Pan Y (2012) High concentration of viscous material pipeline characteristics of the discussion. Mater Sci Forum 704(705):382–388

Zhao LA, Wang TL, Cai RH (2019) Prediction of coal slurry pipeline transportation grading reduction and its influence on pipe transportation parameters. Braz J Chem Eng 36(2):845–853

Chapter 10
Simulation Analysis of Sediment Corrosion of Mixed Transport Pipeline Based on FLUENT

Minglei Xia and Guanglei Lv

Abstract In this paper, ANSYSFLUENT software is used to simulate the corrosion of sediment in the mixed transport pipeline, and explore the corrosion law in the pipeline. The numerical simulation results show that the corrosion rate is affected by the medium velocity. Corrosion caused by sediment accumulation is exacerbated when increased velocity causes the flow pattern to change from laminar to plunger flow. The research results can provide reference and theoretical basis for submarine pipeline corrosion prevention and control, pipeline operation and maintenance.

Keywords Submarine pipeline · Sediment corrosion · Multiphase flow · Numerical simulation

10.1 First Section

In the process of development and production of offshore oil and gas fields, submarine pipeline is an important part of oil and gas gathering and transportation system, which has a safe, economic and reliable mode of oil and gas transportation. However, the submarine pipeline is located in a complex Marine environment, which is difficult to inspect and maintain. Once corrosion leakage accident occurs, it will not only cause huge economic losses, but also lead to pollution and destruction of local environment, and threaten the life safety of Marine operators in serious cases (Xu 2014). In order to ensure the safe operation of submarine pipelines within the design life period, it is necessary to carry out submarine pipeline corrosion simulation and explore the internal corrosion law of pipelines, which has important theoretical and practical significance for extending the service life of pipelines and ensuring the safe development and production of oil and gas fields.

M. Xia (✉) · G. Lv
CNOOC (Tianjin) Pipeline Engineering Technology Ltd, Tianjin 300452, China
e-mail: 1063023182@qq.com

G. Lv
e-mail: lvgl@cnooc.com.cn

H. Xu (ed.), *Proceedings of the 5th International Symposium on Water Resource and Environmental Management*, Environmental Science and Engineering,
https://doi.org/10.1007/978-3-031-31289-2_10

In recent years, some scholars at home and abroad based on the experimental and CFD numerical simulation methods, mostly through changing the fluid solid phase parameters and pipeline parameters (Cui 2018; Sun et al. 2018; Farokhipour and Mansoori 2019; Zhuo et al. 2020; Zolfagharnasab et al. 2021), for optimal structure design standard elbow (Carlos et al. 2016; Chen 2019; Li 2019; Li et al. 2021), The erosion of gas–solid two-phase flow and gas–liquid–solid multiphase flow in single elbow and multi-elbow combined elbow (Jun et al. 2018; Yu et al. 2018; Xie et al. 2020; Othayq et al. 2021) was studied. Among them, the structural optimization of the elbow will increase the processing difficulty and manufacturing cost of the elbow to a certain extent, which is not feasible in practical application. However, there are few researches on the erosion characteristics of multi-bend pipe, and few practical application fields, so the main research is on the erosion inside the single bend pipe. Some scholars have shown that (Zeng 2017), the roughness of solid particles will decrease with the increase of erosion time, and the change of roughness will affect the erosion wear of samples. At present, in the literature on the effect of roughness erosion, most of the studies are about the impact of fluid-to-solid phase parameters, pipeline parameters on the erosion characteristics of targets and the erosion changes of targets with different roughness (Morteza and Hamid 2017; Hao et al. 2017; Liu et al. 2018; Tu 2019). However, there are few researches on the influence of solid particle roughness on erosion, so it is necessary to study the variation law of the erosion characteristics of curved pipe under different particle roughness conditions.

Based on ANSYSFLUENT software, a suitable physical model is established in this paper to numerically simulate the multiphase flow field of the mixed transport pipeline. At the same time, on the basis of the results of flow simulation calculation, the sediment corrosion simulation was carried out in the sea pipe, and the corrosion law of sea pipe under different working conditions was summarized, so as to provide more powerful data reference and theoretical basis for optimizing pipeline design and process technology, pipeline operation and maintenance.

10.2 Mathematical Model

10.2.1 Governing Equation

The continuous phase fluid should follow the three conservation laws of physics, namely, the law of mass conservation, the law of momentum conservation and the law of energy conservation. The mathematical description of the three laws of fluid motion is the basic equations of fluid mechanics. Its specific expression is:

$$\frac{\partial \rho}{\partial t} + \nabla \cdot (\rho \vec{u}) = 0 \tag{10.1}$$

$$\frac{\partial}{\partial t}(\rho u_i) + \frac{\partial}{\partial x_j}(\rho u_i u_j) = -\frac{\partial p}{\partial x_i} + \frac{\partial \tau_{ij}}{\partial x_j} \tag{10.2}$$

$$\frac{\partial}{\partial t}(\rho H) + \frac{\partial}{\partial x_j}(\rho u_j H) = \frac{\partial p}{\partial t} + \frac{\partial}{\partial x_j}(u_i \tau_{ij} - q_j) \tag{10.3}$$

where, u is the velocity vector, m/s; t is time, s; ρ is fluid density, kg/m^3; p is static pressure, Pa; τ_{ij} is the viscous stress tensor, Pa, whose expression is: q_j is the heat flux vector, w/m^2; H is total enthalpy, J/kg.

10.2.2 Turbulence Equation

The selection of turbulence model is very important for numerical simulation. A suitable turbulence model can accelerate the convergence of calculation results and improve the accuracy of simulation. In this paper, the Realizable k-ε turbulence model is adopted. Compared with Standard k-ε model, the Realizable k-ε model adopts a new turbulence viscosity formula, which has a wider application range, especially for the calculation of complex flows with separate flows or secondary flows. Its expression is:

$$\frac{\partial}{\partial t}(\rho k) + \frac{\partial}{\partial x_i}(\rho k u_i) = \frac{\partial}{\partial x_j}(\alpha_k \mu_{eff} \frac{\partial k}{\partial x_j}) + G_k + G_b - \rho\varepsilon - Y_M + S_k \tag{10.4}$$

$$\frac{\partial}{\partial t}(\rho\varepsilon) + \frac{\partial}{\partial x_i}(\rho\varepsilon u_i) = \frac{\partial}{\partial x_j}(\alpha_\varepsilon \mu_{eff} \frac{\partial \varepsilon}{\partial x_j}) + G_{1\varepsilon}\frac{\varepsilon}{k}(G_k + G_{3\varepsilon} G_b) - G_{2\varepsilon}\rho\frac{\varepsilon^2}{k} - R_\varepsilon + S_\varepsilon \tag{10.5}$$

where, G_k is the turbulent kinetic energy generated by the laminar flow velocity gradient, G_b is the turbulent kinetic energy generated by the buoyancy, Y_M is the wave generated by the transition diffusion in compressible flow, C_1, C_2, C_3 are constants, α_k and α_ε are the turbulent Schmidt numbers of equation k and equation ε, S_k and S_ε are user-defined.

10.2.3 Erosion Equation

In the simulation process, particle diameter, quantity and mass flow rate will affect the wall erosion/deposition rate. The rate of abrasion is defined as:

$$R_{\text{erosion}} = \sum_{p=1}^{N_{particles}} \frac{\dot{m}_p C(d_p) f(\alpha) v^{b(v)}}{A_{face}} \tag{10.6}$$

where, C (d_p) is a function of particle diameter; α is the impact Angle (invasion Angle) avoided by particles; f(α) is a function of impact Angle; v is the velocity of the particle relative to the wall surface; b of v is a function of this relative velocity. By default, C = 1, f = 1, and b = 0. Since C, f and b are all defined as the boundary conditions at the wall, rather than the medium properties of certain particles, their default values will not change with different particle materials.

10.3 Hybrid Pipeline Model

10.3.1 Geometric Modeling and Meshing

A CFD simulation model was established based on the mixed transport pipe between an offshore platform. The nominal diameter of the pipe is 273.1 mm and the wall thickness is 11.1 mm. Due to the limitation of the scale of simulation analysis and considering the efficiency of simulation calculation, only the 12 m-long sea pipe was calculated as the research object. Finally, the three-dimensional geometric model of the sea pipe was established, as shown in Fig. 10.1.

The model was divided by the latest Poly-Hexcore mesh technology. Among them, the hexahedral core region can better capture the gradient phenomenon in the calculation region, improve the accuracy and convergence of the simulation calculation, and improve the calculation efficiency while reducing the number of grids.-The near-wall layered grid is used to capture the large gradient calculation area near the wall, which makes the calculation accuracy higher and does not increase the computer burden too much. The polyhedral mesh is used to connect the two parts

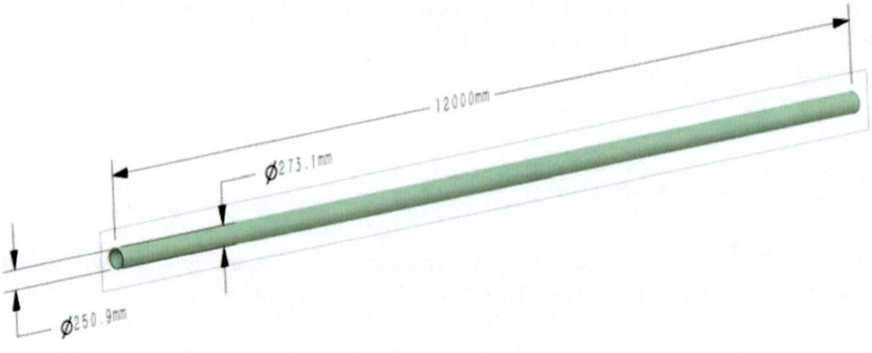

Fig. 10.1 Schematic diagram of three-dimensional geometric model of mixed transport pipeline

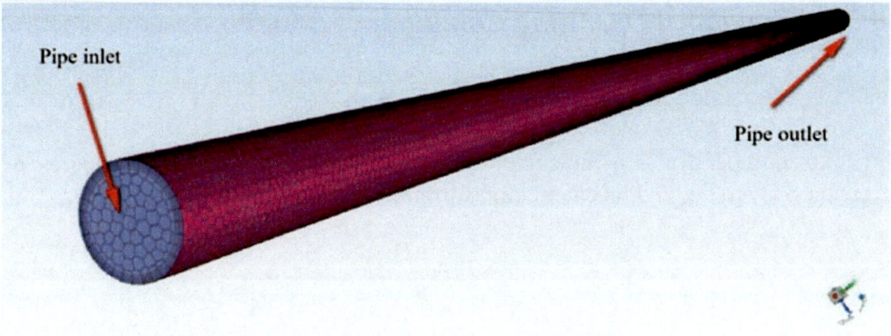

Fig. 10.2 Grid division diagram of mixed transport sea pipeline

of the grid, allowing the entire grid to work at an automated level. Finally, it was determined that the number of calculated grids was 80,388 and the number of nodes was 225,407. The grid was generated, as shown in Fig. 10.2.

10.3.2 Boundary Conditions and Solving Equations

Kerosene, Carbon dioxide and Water are used in the simulation. Multiphase flow model adopts VOF model, which is used to calculate the interface and flow condition of incompatible oil, gas and water. Erosion was simulated using the Erosion/Accretion model in a discrete phase (DPM) model, which can be used to calculate wall abrasion/deposition rates caused by particles.

The inlet of the mixed sea transport pipeline is set as the mass flow inlet, and the mass flow of oil, gas and water three-phase is set respectively. The sea pipe outlet is set as the pressure outlet. The wall of the sea pipe is set as no slip boundary condition. In the DPM model, the particle type is inert. Set surface jet source, particles from the inlet surface into the flow field. Particle diameters are uniformly distributed, with a size of 10^{-6} m. It is assumed that the interaction between particles is ignored in the process of fluid flow, and the collision between medium particles and sea pipe wall is reflected back field. Therefore, escape boundary conditions were selected for DPM of the inlet and outlet surfaces, and reflect boundary conditions were selected for sea pipe wall surfaces.

Coupled algorithm was used to solve the simulation. Considering the convergence and accuracy of the calculation, PRESTO!Format. For the volume fraction, use the "Compressive" format, and for the rest, use the second-order upwind format.

10.3.3 Simulated Operating Condition

10.3.3.1 Reynolds Number Along the Way

Reynolds number Re is a dimensionless parameter used to judge the degree of turbulence in a tube, and its calculation formula is as follows:

$$\mathrm{Re} = \frac{\rho v D}{\mu} \tag{10.7}$$

where, ρ is the fluid density in the pipe, kg/m^3; v is the fluid velocity in the tube, m/s; D is the hydraulic radius, m; μ is the hydrodynamic viscosity in the tube, Pa·s. The greater Re is, the greater the degree of turbulence in the tube. When Re > 2000, the fluid in the tube is considered as turbulence, and the scour effect on the wall is more obvious.

10.3.3.2 Froude Number Along the Way

Froude number Fr is a dimensionless parameter used to characterize the ratio of inertia force to gravity when judging the flow of a liquid with a free surface. The formula is as follows:

$$Fr = \frac{v}{\sqrt{gL}} \tag{10.8}$$

where, v is the fluid velocity in the pipe, m/s; g is the acceleration of gravity, m/s^2; L is the water depth in the pipe, m. The greater the Fr, the greater the fluid inertia and the more turbulent the flow is considered to be.

10.3.3.3 Simulated Location Selection

Based on Reynolds number Re and Froude number Fr, three sections of sea pipe with obvious erosion and high corrosion risk were selected for three-dimensional flow simulation. According to different simulation conditions, the pressure, the mass flow rate of oil, gas and water three-phase, particle velocity and particle mass flow rate are set. See Table 10.1.

Table 10.1 Parameter settings of simulated working conditions

Parameter Pipe section	Inlet temperature (°C)	Outlet temperature (°C)	Inlet pressure (MPa)	Outlet pressure (MPa)	Infusion volume (m^3/d)	Gas transmission ($Nm^3/d*10^4$)	Oil throughput (m^3/d)	Liquid holdup (%)
First paragraph								75
The second paragraph								95
The third paragraph								95

10.4 Simulation Result Analysis

10.4.1 Simulation Results of Multiphase Flow Patterns

Through numerical simulation, the oil–water mixing interface, pipeline profile component distribution diagram (white is water phase, yellow is oil phase, blue is water phase) and pipeline velocity vector diagram of different sea pipeline sections are obtained, as shown below.

(1) Stage 1 Marine Pipe (Liquid Holdup 75%) (Figs. 10.3, 10.4 and 10.5)

(2) Stage 2 Marine Pipe (liquid holding rate 95%) (Figs. 10.6, 10.7 and 10.8)

(3) Stage 3 Marine Pipe (liquid holding rate 95%) (Figs. 10.9, 10.10 and 10.11).

According to the calculation of Re and Fr along the three different positions of the sea pipe, combined with the analysis of the three-dimensional simulation results, it can be concluded that the flow inside the first two sections of the sea pipe is the plunger flow. Under different liquid holding rates, the mixing of oil, gas and water in the pipe is relatively complicated. The flow inside the third section of the pipe is laminar flow. The mixing of oil, gas and water is very stable, and the interlayer interface is clear and the relative position is unchanged.

In fact, in the simulation process, the flow situation is transient and cannot get a stable result. Therefore, the analysis results of the three typical positions of sea pipes above are the flow patterns at a certain time, as well as the corresponding component distribution and velocity vector.

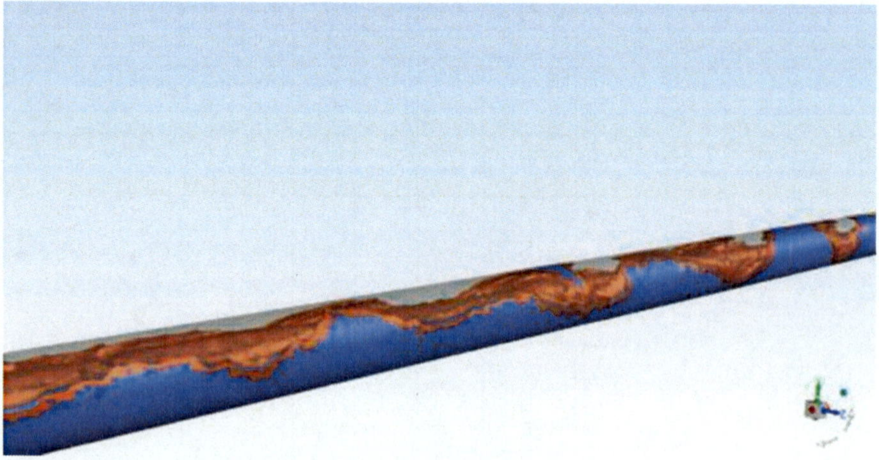

Fig. 10.3 Oil–water mixing interface

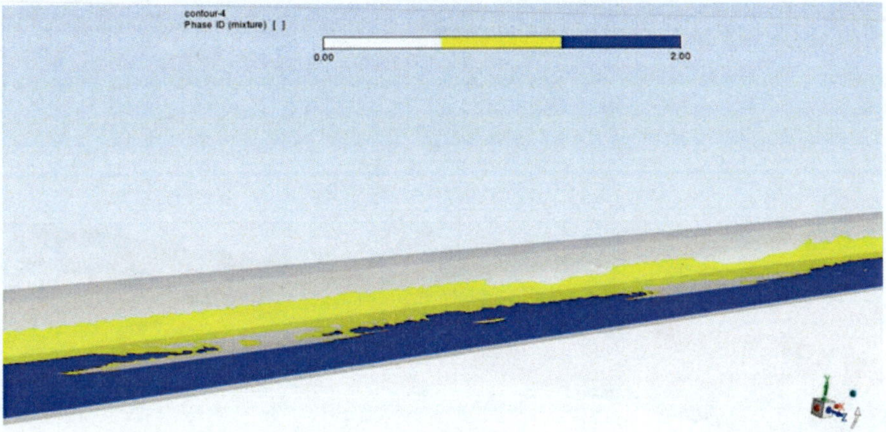

Fig. 10.4 Composition distribution of pipeline profile

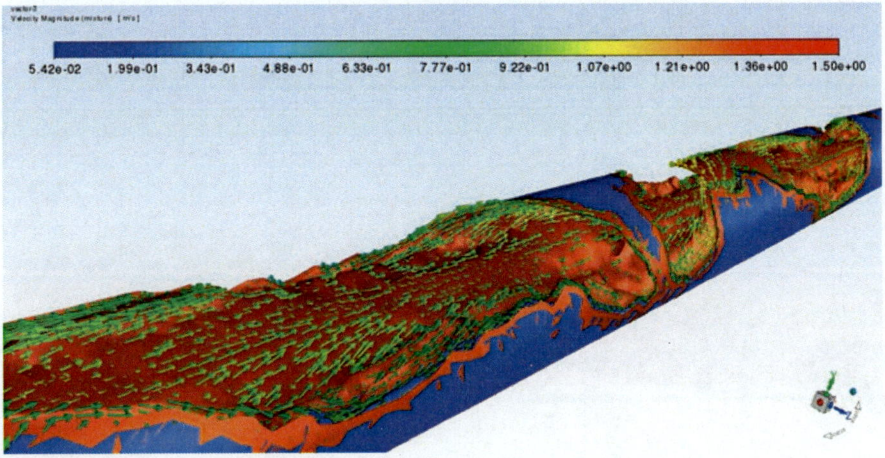

Fig. 10.5 Velocity vector diagram

10.4.2 Simulation Results of Marine Pipe Corrosion

The formation of solid sediment in the pipe will lead to corrosion of the pipe wall. The distribution of sediment is related to many factors, including velocity distribution of flow field in tube, gravity distribution of solid sediment and turbulence near wall. The distribution of sediments under different working conditions can characterize the corrosion of wall surface to a certain extent. Through CFD simulation, the distribution concentration of particle sediments near the wall is shown in the figure below.

(1) Stage 1 Marine Pipe (Liquid Holdup 75%) (Figs. 10.12 and 10.13)

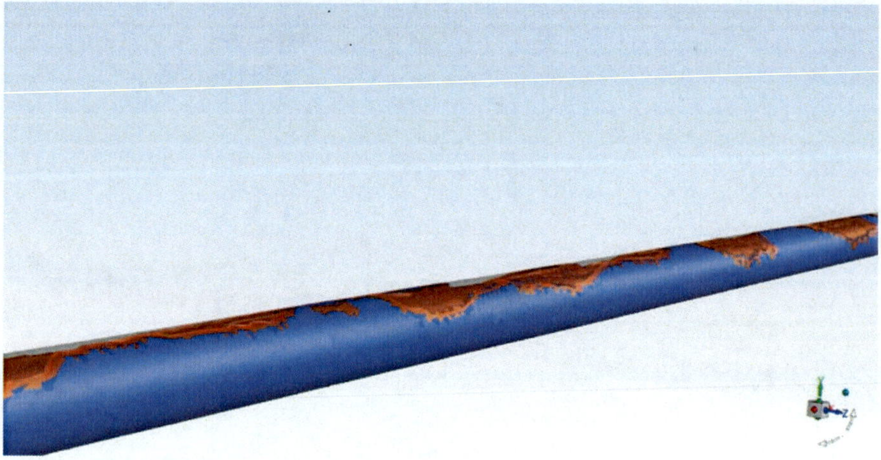

Fig. 10.6 Oil–water mixing interface

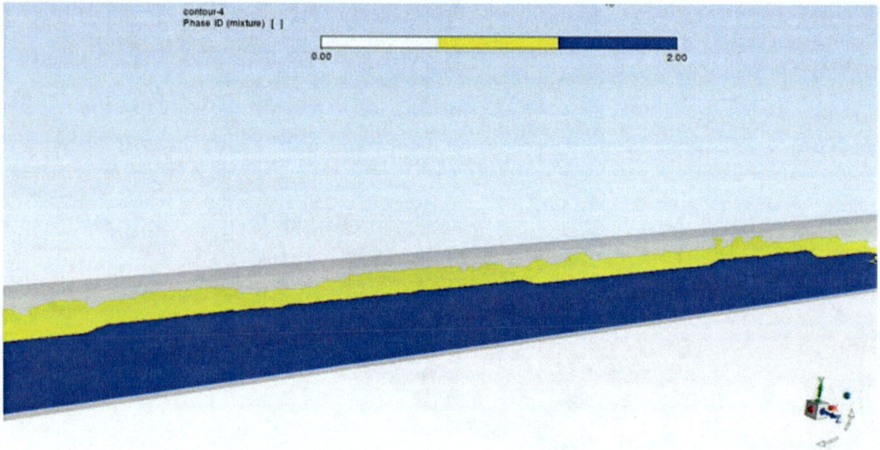

Fig. 10.7 Composition distribution of pipeline profile

(2) Stage 2 Marine Pipe (liquid holding rate 95%) (Figs. 10.14 and 10.15)

(3) Stage 3 Marine Pipe (liquid holding rate 95%) (Figs. 10.16 and 10.17).

According to the analysis of the three-dimensional simulation results, the sediment distribution inside the first two sections of the sea pipe is relatively complex, with large sediment concentration and obvious corrosion. The internal flow in the third section is laminar flow, and corrosion still exists, but it is significantly improved compared with the plunger flow in the first two sections. At the same time, the concentration of sediment at the bottom of laminar flow (order of 10^{-7}) is much

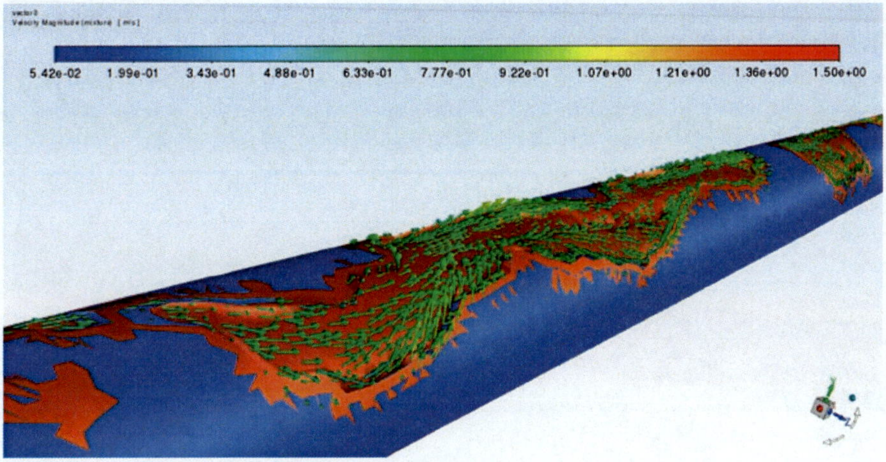

Fig. 10.8 Pipeline velocity vector diagram

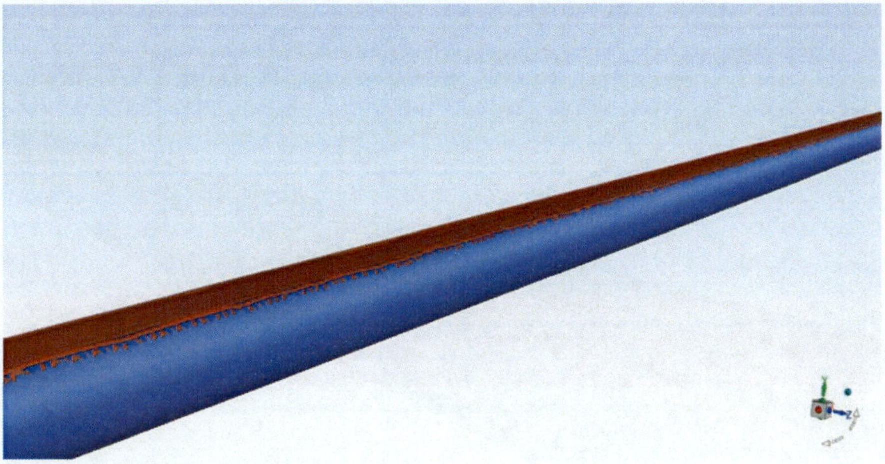

Fig. 10.9 Oil–water mixing interface

smaller than that of the plunger flow deposition (order of 10^{-5}), so it can be considered that the corrosion in laminar flow area is small.

10.5 Prevention and Control Measures

According to the numerical simulation results, medium velocity is one of the main factors affecting sediment corrosion. Therefore, in order to reduce the corrosion of

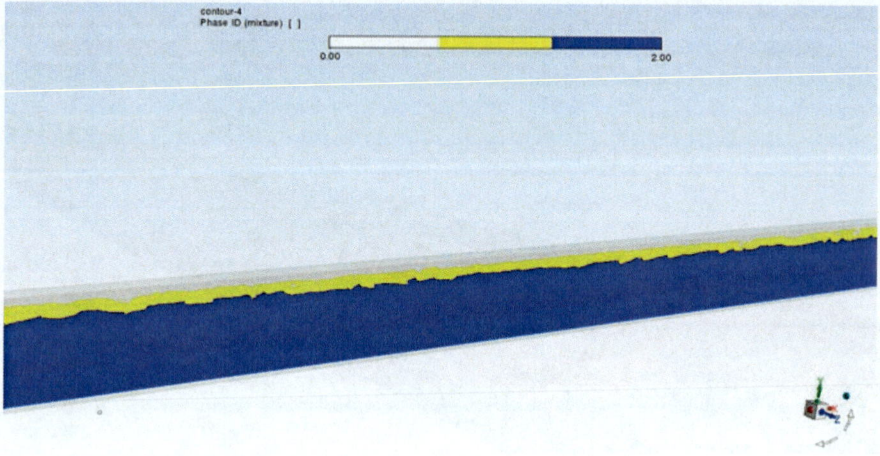

Fig. 10.10 Pipeline profile composition distribution diagram

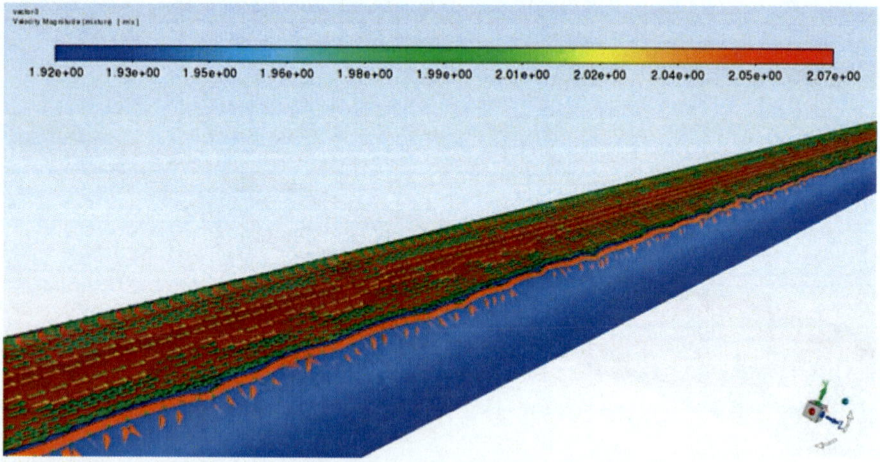

Fig. 10.11 Vector diagram of pipeline velocity distribution

submarine pipelines during operation and extend their service life, the following anti-corrosion measures are proposed:

(1) Control the flow rate of pipe transport medium within the minimum corrosion range, the lower limit of flow rate should make impurities suspended in the pipe transport medium, so that the accumulation of abrasive particles and sediment in the pipeline is the least.
(2) Pigging operation should be carried out regularly to clean the sediment in the pipeline.

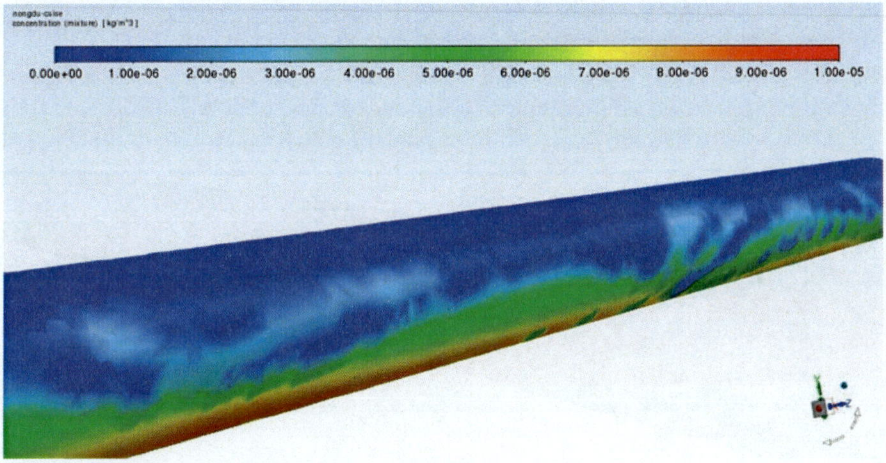

Fig. 10.12 Corrosion sediment concentration profile

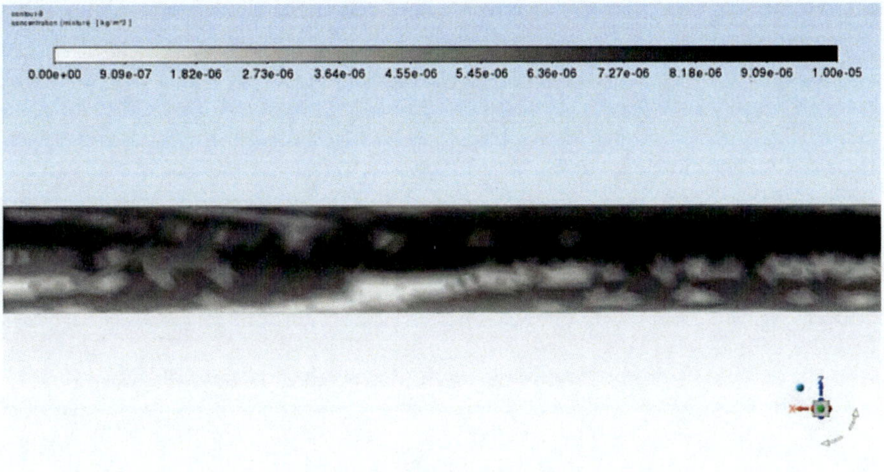

Fig. 10.13 Pipeline profile concentration distribution diagram

(3) Simulation software is used to simulate the high-risk corrosion parts of the sea pipe, and effective protection technology is adopted for the high-risk parts, such as adjusting operation parameters to optimize the process flow, coating protection, selecting appropriate corrosion resistant materials, etc., and key monitoring is implemented.

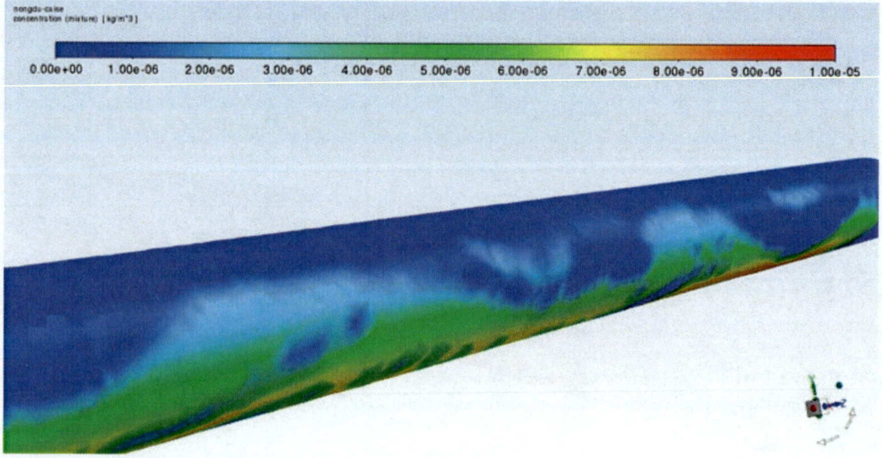

Fig. 10.14 Corrosion sediment concentration profile

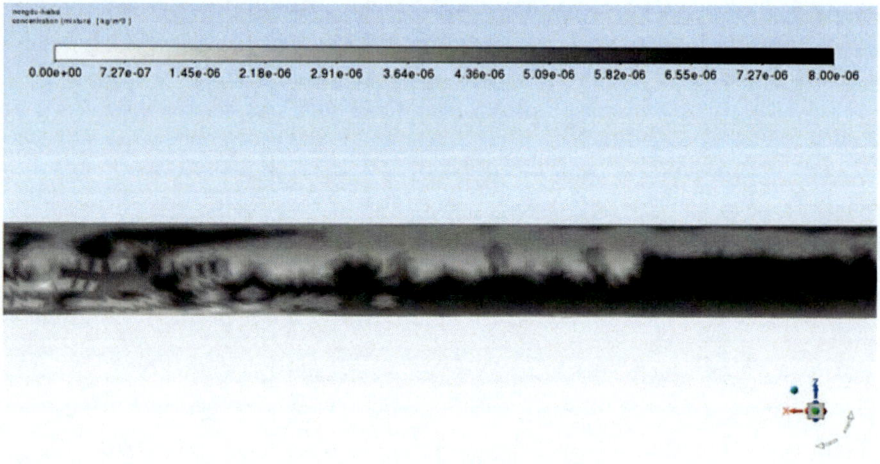

Fig. 10.15 Pipeline profile concentration distribution diagram

10.6 Conclusion

There are various sources of corrosion in sea pipes and different corrosion effects. This paper only simulates the sediment corrosion of submarine pipeline without considering the influence of chemical corrosion on pipeline comprehensively. cThrough the analysis of sediment corrosion under different working conditions during the flow of submarine pipeline, the following conclusions are drawn:

(1) Aiming at the internal corrosion problem of submarine pipeline, ANSYS-FLUENT was used to establish an appropriate physical model for the high-risk

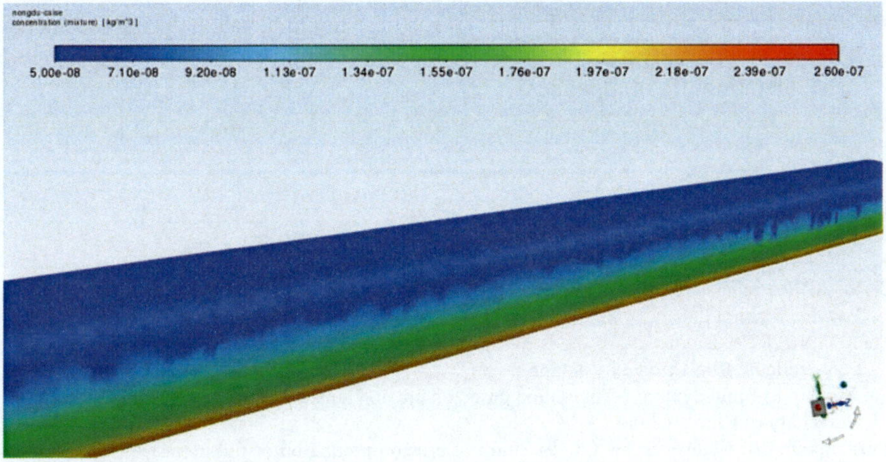

Fig. 10.16 Corrosion sediment concentration profile

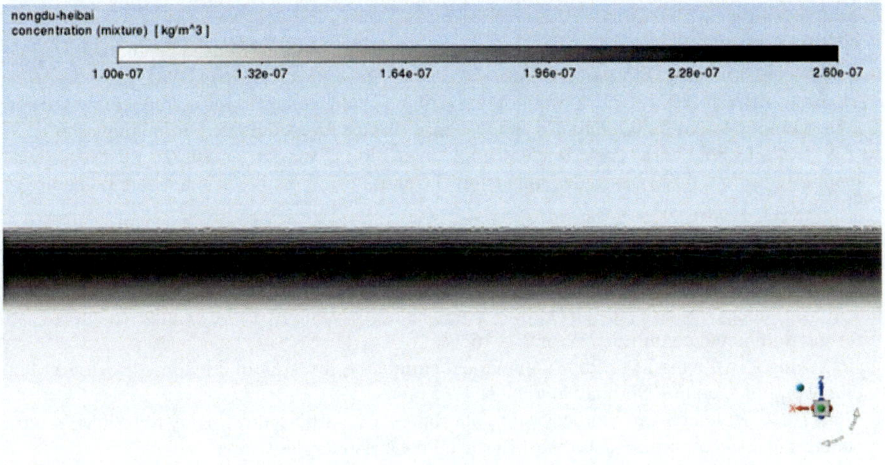

Fig. 10.17 Pipeline profile concentration distribution diagram

positions of Marine pipeline corrosion, and internal corrosion prediction was carried out based on the actual working conditions of Marine pipeline, so as to analyze and evaluate the corrosion development trend of pipeline high corrosion risk during the operation of target pipeline.

(2) The numerical simulation results show that the corrosion rate is affected by the medium velocity. Corrosion caused by sediment accumulation is exacerbated when increased velocity causes the flow pattern to change from laminar to plunger flow. Higher flow rates often result in higher corrosion rates.

(3) Appropriate prevention and control measures should be taken during the operation of submarine pipelines to reduce the risk of sediment corrosion and extend the operation life of pipelines.

References

Carlos ARD, Francisco JDS, Vinicius FDS (2016) Mitigating elbow erosion with a vortex chamber. Powder Technol 288

Chen WN (2019) Bionic study on body surface erosion resistance of semi-desert Parabuthus transvaalicus. Jilin University, China

Cui ZZ (2018) Fluid dynamics simulation based on erosion leakage of oil and gas pipelines. North University of China, China

Farokhipour ZA, Mansoori AR (2019) Study of erosion prediction of turbulent gas-solid flow in plugged tees via CFD-DEM. Powder Technol 352

Hao YH, Liu YL, Xing YM (2017) Erosion wear surface morphology and roughness of steel structure galvanized coating. China Surf Eng 30(1):56–62

Jun Z, McLaury BS, Shirazi SA (2018) Modeling sand fines erosion in elbows mounted in series. Wear 402–403

Li YP (2019) Study on elbow erosion and anti-erosion optimization of gas supercharged station. Southwest Petroleum University, China

Li AJ, Wang ZB, Zhu LY (2021) Design optimization of guide vane for mitigating elbow erosion using computational fluid dynamics and response surface methodology. Particuology 2:6

Liu ZW, Li JF, Li FY (2018) Experimental study on erosion of materials with different heat treatment by ultrafine Al_2O_3 particles impacting. J Central South Univ (Science and Technology) 49(7):1634–1642

Morteza A, Hamid MG (2017) Erosion and erosion-corrosion of Al-brass alloy: effects of jet velocity, sand concentration and impingement angle on surface roughness. Chin J Nonferr Metals (English Edition) 27(11):2371–2380

Othayq M, Haider G, Vieira RE (2021) Effect of distance between two elbows in series on erosion for gas dominated conditions. Wear 2:1–16

Sun ZL, Xing ZH, Zhang MY (2018) Numerical simulation research in erosion-corrosion of 90° elbow pipe. J Liaoning Shihua Univ 38(2):47–51

Tu CY (2019) Research on characteristics of iron oxide particle erosion of a 1000MW ultra supercritical steam turbine. Northeast Electric Power University, China

Xie ZQ, Cao XW, Zhang JN (2020) Study of erosion behavior in elbows mounted in series using computational fluid dynamics method. J Energy Res Technol 142(11):1–10

Xu WX (2014) Analysis and protection of internal corrosion of submarine oil and gas pipeline. Corros Protect 35(5):500–504

Yu W, Qi H, Fei H (2018) Numerical simulation of the erosion characteristics and structure optimization of elbows connection for gas-solid flow. Proc Chin Soc Electr Eng 38(3):832–839

Zeng ZH (2017) Investigation of particle impact on 304SS under the liquid-solid two phase flow. Xiamen University, China

Zhuo K, Qiu YJ, Zhang YD (2020) Influence of particle parameters on erosion wear of gas-solid two-phase flow in pipeline. J Xi'an Shiyou Univ (Natural Science Edition) 35(6):100–106

Zolfagharnasab MH, Salimi M, Zolfagharnasab H (2021) A novel numericalinvestigation of erosion wear over various 90-degree elbow duct sections. Powder Technol 380

Chapter 11
The Comparative Assessments of Growth Status of Sugarcane by TOPSIS Methods and Tissue Elements Distributions on Farmland of Heavily Polluted Environment

Kaixian Zhu, Guiping Xu, Xiaofei Wang, Hexiao Chen, and Shiyang Yan

Abstract The growth status of six sugarcane varieties on heavy metal contaminated field were assessed by six selected indicating traits—percentage of germination, plant height, Brix, stem diameters, millable stalks and yield through analyzing by TOPSIS methods, then elements distributions in sugarcane tissues were analyzed by ICP-MS, An simple algorithm of element distributions in sugarcane leaves and roots was designed to rank the tolerance of sugarcane varieties in heavy metal polluted farmland. The results had indicated the element distributions in sugarcane tissues were highly corresponding to the overall TOPSIS scoring of sugarcane traits.

Keywords Sugarcane · Pollution · TOPSIS · Tolerance

11.1 Introduction

Pollutions is a common concerning issues for environment sustainability, for instance, the contamination of heavy metal in soil, which is persisted for long time and risk inhabitants' living through crops or other exposure routes (Hu et al. 2013; Christou et al. 2014; Li et al. 2014). In less developed countries, economic activities of mining or industries is easy to release heavy metals to environment around as there are little environmental control measures and the polluted land is usually used for growing crops due to shortages of available arable land (Liu et al. 2013, 2014; Anju and Banerjee 2012). To minimize the negative effects of soil heavy metals to human

K. Zhu · H. Chen · S. Yan
Scientific Research Academy of Guangxi Environmental Protection, Nanning 530000, China
e-mail: kaixianzhu@163.com

G. Xu (✉)
Department of Ecology and Environement of Guangxi, Nanning 530028, China
e-mail: xuguiping@126.com

X. Wang
Hubei Polytechnic University, Huangshi 435003, China

H. Xu (ed.), *Proceedings of the 5th International Symposium on Water Resource and Environmental Management*, Environmental Science and Engineering,
https://doi.org/10.1007/978-3-031-31289-2_11

being, one sustainable way is to use contaminated field for planting economical crops not for direct eating by human, for instance, sugarcane used for ethanol production (Xie et al. 2014). The technique for order performance by similarity to ideal solution (TOPSIS) which was firstly develop by Hwang and Yoon (1981), had been widely utilized for Multiple Criteria Decision Making (MCDM) problems in technologies and engineering (Monjezi et al. 2012; Byun and Lee 2005), in recent years, Dynamic TOPSIS methods proposed by Yao et al. (1994) which had been applied to assess different varieties of sugarcanes and other economical crops (Chen and Ma 2006; Shen and Dai 2005; Li et al. 2012; Ma et al. 2013). In this study, scoring the growth status of indicator traits of different sugarcane varieties on highly polluted farmland was performed by the TOPSIS methods firstly, an simple algorithm of combination of biological accumulation coefficient (BAC) and biological transfer coefficient (BTC) figures of each sugarcane variety was then used to scoring the tolerance of each sugarcane variety in heavy metal polluted environments, finally, the scoring of two separate methods were compared to see if the outcomes match each other. Since the scoring systems are on initial stages, each investigated factor (sugarcane variety, trait and element) incorporated only six variables in this investigation.

11.2 Material and Methods

11.2.1 The Trial Land and Sugarcane Growing

The trial field (108.2513N, 25.11759E) is located in Guangxi, China. The land was heavily polluted by water flooded over from an abandoned ore mining site nearby. Trial land was divided into small trial zones of 25 m^2 (5×5 m). 21 sugarcane seedlings were planted on each zone; plants stand away from each other at distances 1×1 m, selected varieties were denoted as TT 27, GT 42, GT 29, LC 03/1137, LC 05/136, GF 98/296 (varieties selected included sub-varieties of the same series). Sugarcane seedlings were grown in duration of 1 year (early April to late March next year) under common sugarcane growing conditions.

11.2.2 Sample Treatment and Analysis

Sample collecting. Six ripe sugarcane plants of each variety were randomly collected in each trial zone, three of them were collected with roots, including portions of root plus soil, stems and leaves, and the other three ones were only collected for the above-ground parts.

Tests of heavy metals in sugarcane and soil. Sugarcane samples to be tested were washed by tap water firstly, and then by purified water, chopped into sections of roots, stems and leaves respectively, the sections were then heated at 105 °C for a

Table 11.1 Conditions of microwave digestion (soil samples)

Warming duration (min)	Digestion temperature (°C)	Digest duration (min)
7	Room–120	3
5	120–160	3
5	160–190	25

short time, dried at 72 °C to a constant weight, grinded into power for analysis. The soil samples were dried at ventilated space, grinded into powder and filtered by 100 mesh sieves for analysis. 0.1 g of soil portion was accurately weighed and placed in a Micro-wave tube, slowly added 6.0 mL Nitric acid, 2.0 mL Hydrochloric acid, 2.0 mL Hydrofluoric acid, the reaction mixture was left in a still condition for the reactions reached an equilibrium. Then Micro-wave tubes were closed and heated in Microwave Digestion Oven as per Conditions listed in Table 11.1. The tubes were standing for 15–30 min cooling after the digestion then the mixture was then transferred to PTFE crucible and heating at 110–120 °C on hotplates to let the acid evaporated, the mixture was left cooled and then transferred into 50 mL volumetric flasks with Milli-Q water added to the standard line. Blank samples were reproduced with the same reagents and experimental conditions. Prepared samples were diluted to suitable volumes in case of need and then were tested by the methods of ICP-MS.

0.5 g of sugarcane portion was accurately weighed and placed in a Micro-wave tube, slowly added 8.0 mL hydrofluoric acid; the reaction mixture was left in a still condition for the reactions reached equilibrium. Then Micro-wave tubes were closed and heated in Microwave Digestion Oven as per Conditions listed in Table 11.2. The tubes were standing for 15–30 min cooling after the digestion then the mixture was then transferred to PTFE crucible and heating at 110–120 °C on hotplates to let the acid evaporated, the mixture was left cooled and then transferred into 50 mL volumetric flasks with Milli-Q water added to the standard line. Blank samples were reproduced with the same reagents and experimental conditions. Prepared samples were diluted to suitable volumes in case of need and then were tested by the methods of ICP-MS.

Calculations of Biological Accumulation and Transfer Coefficients. The figure of Biological Accumulation Coefficient (BAC) is an indicator that how able a plant assimilates benefit elements or tolerates harm elements; it is calculated by the ratios of accurate element levels measured in plant tissues to element levels in soil, which is expressed by the equation below:

Table 11.2 Conditions of microwave digestion (sugarcane samples)

Warming duration (min)	Digestion temperature (°C)	Digest duration (min)
7	Room–120	3
5	120–160	3
5	160–190	25

$$BAC = \frac{C_p}{C_s} \tag{11.1}$$

where: BAC—Biological Accumulation Coefficient, C_p—elements level (concentration) of Plant tissues; C_s—elements level (concentration) of soil.

The figure of Biological Transfer Coefficient (BTC) is an indicator that how elements migrate from plants' root to their up-ground organs, it is calculated by the ratios of accurate elements level measured in above-ground plant tissues to element levels in roots, which is expressed by the equation below:

$$BTC = \frac{C_a}{C_r} \tag{11.2}$$

where: BTC—Biological Transfer Coefficient, C_a—elements level (concentration) of aboveground organs; C_r—elements level (concentration) of roots.

11.3 Results and Discussion

11.3.1 The Growth Traits Analysis of Different Sugarcane Varieties

Six sugarcane traits were selected for indicating positive growth, For each variety, the data of percentage of germination (PG) and plant heights (PH, unit: m) were recorded and the Brix (measured by hand refractometer), stem diameters (SD, unit: cm), millable stalks (MS, estimated number of millable stalks/Chinese acre) and yield (estimated kg/Chinese acre) were measured at the time of harvest (see Table 11.3). The six traits were used to score the growth status of sugarcane on contaminated field, weights of each traits was assigned empirically as per growing expertise (weight of PG = 0.2, PH = 0.1, Brix = 0.2, SD = 0.2, MS = 0.1, Yield = 0.2), the figures of traits were used to calculate Decision Matrix according to common TOPSIS procedures (Monjezi et al. 2012; Zhao et al. 2008).

Table 11.3 Figures of six traits of six sugarcane varieties on harvest

Variety	PG (%)	PH	SD	Brix	MS	Yield
TT 27	42.9	2.01	2.46	22.4	4266.7	3474.0
GT 42	76.2	2.40	2.56	23.0	5466.7	5903.2
GT 29	42.9	1.71	2.01	22.9	7200.0	3201.7
LC 03/1137	71.4	2.22	2.54	23.6	4800.0	4675.7
LC 05/136	61.9	1.97	2.55	23.1	5333.4	4537.1
GF 98/296	52.4	2.17	2.47	22.7	4533.4	4052.5

The data of each trait was then normalized into the form Z_{ij}, which is calculated as:

$$Z_{ij} = Y_{ij}/Y_i\max, i = 1, 2, 3 \ldots 6; j = 1, 2, 3 \ldots 6$$

where Y_i max = max (Y_{ij}) as long as all traits studied here are positive indicators of sugarcane growth. The evaluation matrix (Z_{ij}) m × n (m = 6, n = 6) is expressed as:

$$Z = \begin{bmatrix} 0.56\ 0.84\ 0.96\ 0.95\ 0.59\ 0.59 \\ 1.00\ 1.00\ 1.00\ 0.98\ 0.76\ 1.00 \\ 0.56\ 0.71\ 0.78\ 0.97\ 1.00\ 0.54 \\ 0.94\ 0.92\ 0.99\ 1.00\ 0.67\ 0.79 \\ 0.81\ 0.82\ 1.00\ 0.98\ 0.74\ 0.77 \\ 0.69\ 0.90\ 0.97\ 0.96\ 0.63\ 0.69 \end{bmatrix}$$

The matrix Z is transformed into the normalized decision matrix R with r_{ij} as normalized values:

$$r_{ij} = \frac{Z_{ij}}{\sqrt{\sum_{k=1}^{m} Z_{ij}^2}}, i = 1, 2, 3 \ldots m; j = 1, 2, 3 \ldots n$$

The matrix R is expressed as:

$$R = \begin{bmatrix} r_{11} & \cdots & r_{16} \\ \vdots & \ddots & \vdots \\ r_{61} & \cdots & r_{66} \end{bmatrix} = \begin{bmatrix} 0.29\ 0.39\ 0.41\ 0.40\ 0.32\ 0.32 \\ 0.52\ 0.47\ 0.43\ 0.41\ 0.42\ 0.55 \\ 0.29\ 0.33\ 0.33\ 0.41\ 0.55\ 0.30 \\ 0.49\ 0.43\ 0.42\ 0.42\ 0.37\ 0.43 \\ 0.42\ 0.38\ 0.43\ 0.41\ 0.41\ 0.42 \\ 0.36\ 0.42\ 0.42\ 0.40\ 0.35\ 0.38 \end{bmatrix}$$

Weight coefficient W_j of each trait was allocated as:

$$W_j = \{0.20.10.20.20.10.2\}, W_j \in (0, 1), \sum_{j=1}^{6} W_j = 1$$

And the criteria weighted matrix is calculated as:

$$W = \begin{bmatrix} w_1 & \cdots & 0 \\ \vdots & \ddots & \vdots \\ 0 & \cdots & w_6 \end{bmatrix}$$

The weighted normalized decision matrix D ($d_{ij} = w_j\ r_{ij}$) is calculated as:

Table 11.4 Six traits of six sugarcane varieties scored by TOPSIS

Variety	PG	PH	SD	Brix	MS	Yield	Score (sum)
TT 27	0.059	0.039	0.082	0.080	0.032	0.065	0.357
GT 42	0.105	0.047	0.086	0.082	0.042	0.110	0.471
GT 29	0.059	0.033	0.067	0.081	0.055	0.059	0.354
LC03/1137	0.098	0.043	0.085	0.084	0.037	0.087	0.434
LC05/136	0.085	0.038	0.086	0.082	0.041	0.084	0.416
GF98/296	0.072	0.042	0.083	0.081	0.035	0.076	0.388

$$\mathrm{D} = \mathrm{W} \times \mathrm{R}$$

$$\mathrm{D} = \begin{bmatrix} 0.29 * 0.2 & 0.39 * 0.1 & 0.41 * 0.2 & 0.40 * 0.2 & 0.32 * 0.1 & 0.32 * 0.2 \\ 0.52 * 0.2 & 0.47 * 0.1 & 0.43 * 0.2 & 0.41 * 0.2 & 0.42 * 0.1 & 0.55 * 0.2 \\ 0.29 * 0.2 & 0.33 * 0.1 & 0.33 * 0.2 & 0.41 * 0.2 & 0.55 * 0.1 & 0.30 * 0.2 \\ 0.49 * 0.2 & 0.43 * 0.1 & 0.42 * 0.2 & 0.42 * 0.2 & 0.37 * 0.1 & 0.43 * 0.2 \\ 0.42 * 0.2 & 0.38 * 0.1 & 0.43 * 0.2 & 0.41 * 0.2 & 0.41 * 0.1 & 0.42 * 0.2 \\ 0.36 * 0.2 & 0.42 * 0.1 & 0.42 * 0.2 & 0.40 * 0.2 & 0.35 * 0.1 & 0.38 * 0.2 \end{bmatrix}$$

The score (sum of d values of 6 traits in matrix D) calculated by TOPSIS methods for each variety of sugarcane is summarized in Table 11.4.

Through the scoring by TOPSIS methods to sugarcane variety growth traits, it was found that ranking of growth quality for sugarcanes on polluted field was GT 42 > LC 03/1137 > LC 05/136 > GF 98/296 > TT 27 > GT 29. The score of TOPSIS is used as a measurement of growth status of six varieties.

11.3.2 The Test Results of Elements in Soil

The content of soil elements determined by ICP-MS was listed in Table 11.5. The results showed that most elements tested are much higher (particularly Pb) compared to the background levels of soils in Guangxi (BL) (Institute of Environmental Science of Guangxi Zhuang Autonomous Region 1992). Not available information about background level of soil Fe in Guangxi was identified in the range of publishing documents; however, Fe is obviously high by the tests.

Table 11.5 The soil elements tested by ICP-MS

Heavy metals	mg/kg	BL
Cu	28.88 ± 3.64	18.84
Zn	450.53 ± 75.65	56.26
Pb	849.43 ± 192.13	17.63
Cd	0.48 ± 0.03	0.1015
As	62.51 ± 7.48	11.26
Fe	59317.87 ± 7997.79	No data

11.3.3 The Test Results of Metal Elements in Sugarcane Parts

The content of elements in six sugarcane varieties is tested by ICP-MS; the metal concentrations in three different organs—root, stem and leaves were listed in Table 11.6. The level of Fe was found overwhelming high compared to the others.

11.3.4 The Analysis of Elements Distribution Patterns in Sugarcane Varieties

Finally, the distribution patterns of elements in sugarcane varieties is analyzed by algorithm designed, 6 elements were categorized as 3 beneficial (+) elements—Zn, Fe and Cu which act as essential nutrient elements for the growth of plants; 3 harm (−) elements—As, Pb and Cd which play not recognized biological role in plant and increased the burden of plant tissue developments.

BAC coefficients (Eq. 11.1) of roots and BTC coefficients (Eq. 11.2) of leaves were calculated by Eqs. 11.1 and 11.2 described in Sect. 2.2. The reason these two figures were selected is that most elements were identified tending to accumulated in roots and leaves, and the concentrations of elements in leaves is relied on their concentrations in roots rather than soil. The two coefficients are combined as C values (see Table 11.7), C values of beneficial elements were summed and minus the C values of harm elements to calculate scores of C for particular varieties as described below:

$$\text{Score of C} = C_{Zn} + C_{Fe} + C_{Cu} - C_{As} - C_{Pb} - C_{Cd} \quad (11.3)$$

where C value is the addition of two coefficients (BAC of roots and BTC of leaves), BAC values, BTC values and C values of each element in each variety, also Score of C for all varieties are summarized in Table 11.7. Values of TOPSIS and Score of C for 6 sugarcane varieties are depicted in Fig. 11.1.

The two score system had ranked the six sugarcane varieties in the order GT 42 > LC 03/1137 > LC 05/136 > GF 98/296 > TT 27 > GT 29, since the score of C is

Table 11.6 The metal contents in three sugarcane organs tested by ICP-MS (mg/kg)

Variety	Organ	Zn	As	Pb	Fe	Cu	Cd
TT 27	Root	73.76 ± 20.34	4.10 ± 2.60	94.32 ± 37.49	4621.4 ± 478.39	9.38 ± 1.51	0.39 ± 0.03
	Stem	42.82 ± 10.63	0.22 ± 0.20	5.46 ± 1.48	70.61 ± 7.04	2.65 ± 0.34	0.08 ± 0.02
	Leaf	50.19 ± 12.25	0.59 ± 0.16	12.97 ± 3.90	410.68 ± 50.75	3.47 ± 0.78	0.04 ± 0.01
GT 42	Root	50.63 ± 15.01	3.88 ± 1.93	80.15 ± 36.27	5042.9 ± 662.6	5.48 ± 0.59	0.31 ± 0.04
	Stem	46.33 ± 10.13	0.15 ± 0.08	4.38 ± 1.27	84.18 ± 7.7	2.58 ± 0.52	0.11 ± 0.04
	Leaf	63.10 ± 9.66	0.79 ± 0.25	17.59 ± 3.94	545.68 ± 38.98	4.18 ± 0.96	0.04 ± 0.01
GT 29	Root	130.72 ± 48.95	10.20 ± 9.54	219.8 ± 147.2	5560.5 ± 731.69	9.28 ± 1.25	0.50 ± 0.04
	Stem	57.35 ± 18.37	0.24 ± 0.21	8.03 ± 2.02	98.99 ± 8.56	2.26 ± 0.39	0.20 ± 0.07
	Leaf	65.71 ± 9.56	0.26 ± 0.08	13.32 ± 5.45	158.1 ± 38.98	3.78 ± 0.4	0.03 ± 0.01
LC 03/1137	Root	75.23 ± 25.93	6.73 ± 3.21	125.74 ± 51.71	7614.1 ± 527.3	7.76 ± 0.56	0.22 ± 0.05
	Stem	55.3 ± 13.85	0.11 ± 0.08	5.09 ± 1.26	71.6 ± 8.2	3.13 ± 0.56	0.09 ± 0.04
	Leaf	47.69 ± 6.79	0.23 ± 0.08	10.15 ± 2.75	154.58 ± 37.53	3.24 ± 0.31	0.03 ± 0.01
LC 05/136	Root	87.19 ± 37.4	4.55 ± 2.12	93.48 ± 35.73	6576.4 ± 567.4	7.73 ± 0.96	0.21 ± 0.08
	Stem	57.62 ± 15.74	0.23 ± 0.07	3.98 ± 0.44	69.56 ± 9.32	3.99 ± 0.54	0.16 ± 0.08
	Leaf	50.04 ± 8.9	0.29 ± 0.09	14.66 ± 4.46	172.69 ± 39.54	3.69 ± 0.58	0.04 ± 0.01
GF 98/296	Root	89.98 ± 17.86	11.59 ± 5.4	218.42 ± 93.27	9683.2 ± 580.7	8.91 ± 1.17	0.23 ± 0.05
	Stem	39.59 ± 7.06	0.12 ± 0.07	4.89 ± 1.34	71.43 ± 8.90	2.68 ± 0.76	0.08 ± 0.03
	Leaf	41.48 ± 7.28	0.23 ± 0.06	8.44 ± 2.46	126.87 ± 20.15	3.14 ± 0.40	0.03 ± 0.01

Table 11.7 The score of C for 6 elements in 6 varieties

Variety		Zn(+)	As(−)	Pb(−)	Fe(+)	Cu(+)	Cd(−)	Score of C
TT 27	BAC(R)	0.16	0.07	0.11	0.08	0.32	0.81	0.33
	BTC(L)	0.68	0.14	0.14	0.09	0.37	0.10	
	C	0.84	0.21	0.25	0.17	0.69	0.92	
GT 42	BAC(R)	0.11	0.06	0.09	0.09	0.19	0.65	1.15
	BTC(L)	1.25	0.20	0.22	0.11	0.76	0.13	
	C	1.36	0.27	0.31	0.19	0.95	0.77	
GT 29	BAC(R)	0.29	0.16	0.26	0.09	0.32	1.04	0.03
	BTC(L)	0.50	0.03	0.06	0.03	0.41	0.06	
	C	0.79	0.19	0.32	0.12	0.73	1.10	
LC 03/1137	BAC(R)	0.17	0.11	0.15	0.13	0.27	0.46	0.67
	BTC(L)	0.63	0.03	0.08	0.02	0.42	0.14	
	C	0.80	0.14	0.23	0.15	0.69	0.59	
LC 05/136	BAC(R)	0.19	0.07	0.11	0.11	0.27	0.44	0.62
	BTC(L)	0.57	0.06	0.16	0.03	0.48	0.19	
	C	0.77	0.14	0.27	0.14	0.75	0.63	
GF 98/296	BAC(R)	0.20	0.19	0.26	0.16	0.31	0.48	0.39
	BTC(L)	0.46	0.02	0.04	0.01	0.35	0.13	
	C	0.66	0.21	0.30	0.18	0.66	0.61	

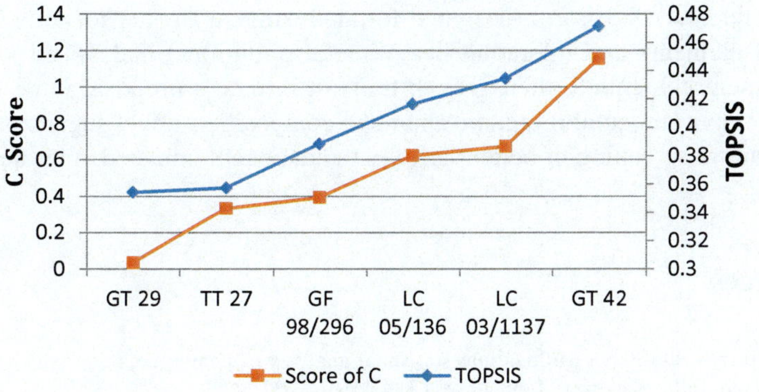

Fig. 11.1 Scoring comparison of 6 varieties of sugarcanes by TOPSIS and score of C

designed to analysis the ability of a particular sugarcane variety to uptake beneficial elements and tolerate harmful elements, it justified the distribution of elements in sugarcane play an important role in its growth status.

However, the scoring systems still have limitations on this stage, firstly, the scope of this research only cover 6 varieties of sugarcane (in this study, only varieties which had been developed more than one sub-varieties had been chosen), in further studies, the sampling scope could be extended to more varieties, more traits of sugarcane (maybe some negative indicator traits) and more other elements could be included into the scoring system to improve the efficacy of the assessment in the future. Since the weight of coefficients of six traits in TOPSIS analysis were assigned based on routine experiences, it had been noticed that the variance of TOPSIS scores had not been varied obviously, in contrast, there is no particular coefficient of weight to be assigned to the elements on this stage that the graphs of C score had been observed changed dramatically, presumably more sugarcane varieties, more traits and more elements would be included in the future research (as an initial stage, only six variable for each factor had been involved in this study), thus to improve both scoring systems, the algorithms should be optimized to meet more high-quality standards of scientific assessment for sugarcane developments, ideally with more statistical approaches particularly in weight coefficient attributions to both traits and elements.

11.4 Conclusion

In the assessment with sugarcane growth, six positive indicator traits of sugarcane growth status were investigated, the situations of six varieties were ranked with TOPSIS analysis, the ranking was justified by the analysis of elements distribution pattern and an algorithm designed to score the tolerance of sugarcane varieties on heavy metal contaminated lands. Overall, the comparative assessment of TOPSIS methods and C score system (designed for analyzing sugarcane for utilizing environmental elements and tolerating heavy metal pollutions) had linked sugarcane tissue element contents to their growth traits, it initiate a proposal route to investigate how environmental inorganic chemical composition affect sugarcane growth and help future screening of better varieties to adapt contaminated environments.

References

Anju M, Banerjee D (2012) Multivariate statistical analysis of heavy metals in soils of a Pb-Zn mining area, India. Environ Monit Assess 184:4191–4206

Byun H, Lee K (2005) A decision support system for the selection of a rapid prototyping process using the modified TOPSIS method. Int J Adv Manuf Technol 26(11–12):1338–1347

Chen F, Ma C (2006) Evaluation on silage new corn cultivar by DTOPSIS analytical method. J Hebei Agric Sci 10:48–51

Christou A, Eliadou E, Michael C, Hapeshi E, Fatta-Kassinos D (2014) Assessment of long-term wastewater irrigation impacts on the soil geochemical properties and the bioaccumulation of heavy metals to the agricultural products. Environ Monit Assess 186(8):4857–4870

Hu Y, Liu X, Bai J, Shih K, Zeng E, Chen F (2013) Assessing heavy metal pollution in the surface soils of a region that had undergone three decades of intense industrialization and urbanization. Environ Sci Pollut Res Int 20(9):6150–6159

Hwang C, Yoon K (1981) Multiple attribute decision making: methods and applications. Springer-Verlag, New York

Institute of Environmental Science of Guangxi Zhuang Autonomous Region (1992) Atlas of soil environmental background values of Guangxi Zhuang Autonomous Region. Chengdu Cartographic Publishing House, Chengdu

Li Y, Li S, Wu J, Meng Z, Guo F (2012) Application of DTOPSIS and grey relational analysis in evaluating newly introduced flue–cured tobacco varieties. Acta Tabacaria Sin 18:35–40

Li Z, Ma Z, Kuijp T, Yuan Z, Huang L (2014) A review of soil heavy metal pollution from mines in China: pollution and health risk assessment. Sci Total Environ 468:843–853

Liu G, Tao L, Liu X, Hou J, Wang A, Li R (2013) Heavy metal speciation and pollution of agricultural soils along Jishui River in non-ferrous metal mine area in Jiangxi Province, China. J Geochem Explor 132:156–163

Liu G, Yu Y, Hou J, Xue W, Liu X, Liu Y, Wang W, Alsaedi A, Hayat T, Liu Z (2014) An ecological risk assessment of heavy metal pollution of the agricultural ecosystem near a lead-acid battery factory. Ecol Ind 47:210–218

Ma W, Guo Q, Li Y, Li H, Zhu R, Luo X, Li M, Wei C, Tang J, Qiu W (2013) Comprehensive evaluation of new sugarcane lines by DTOPSIS method. Sugar Crop China 1:20–22

Monjezi M, Dehghani H, Singh T, Sayadi A, Gholinejad A (2012) Application of TOPSIS method for selecting the most appropriate blast design. Arab J Geosci 5:95–101

Shen X, Dai H (2005) Application of DTOPSIS in evaluation of new tomato varieties. China Veg 5:4–6

Xie J, Weng Q, Ye G, Luo S, Zhu R, Zhang A, Chen X, Lin C (2014) Bioethanol production from sugarcane grown in heavy metal-contaminated soils. BioResources 9(2):2509–2520

Yao X, Zhu Y, Zhang P (1994) Method for multi-objectives evaluation on economics in rural area. Syst Sci Compr Stud Agric 10:23–26

Zhao P, Wu C, Chen X, Liu J, Zhao J, Yang H (2008) Application of DTOPSIS on evaluating sugarcane lines. Sugar Crops China 1:43–46

Chapter 12
Analysis of Water Demand Forecasting System Based on GIS Water Diversion in Yellow River Irrigation Area

Wei Wang and Zhe Liu

Abstract The Yellow River irrigation area in the lower reaches of the Yellow River has excellent natural conditions and is an important food and cotton production base in China. Monitoring soil moisture in the current period is of great significance for ensuring local agricultural production. Based on this, in order to help the relevant personnel use GIS technology to carry out water demand forecasting for the Yellow River irrigation area, this paper starts from the characteristics of GIS-diversion Yellow River irrigation area water demand forecasting system, describes and explains in detail the operation conditions of system model, system content, system structure, system development, system database and system application.

Keywords Water demand forecasting system · GIS water diversion · Yellow river irrigation area

12.1 Introduction

Although the water resources of Yellow River account for a small proportion of the country's total water resources, the areas it passes through play a very important role in China's agricultural field and have a profound impact on ensuring the agricultural production on both sides of the Yellow River, however, with the continuous increase in the amount of irrigation, the contradiction of water consumption has become increasingly prominent, therefore, for water consumption, it is necessary to adopt a unified scheduling method to rationally arrange agricultural water consumption, so as to ensure that the residents living water consumption and industrial water consumption are not affected, in this aspect, GIS technology can play an important role (Wang et al. 2015). In full consider of the factors such as initial soil moisture content, soil moisture regression process and water consumption per irrigation of the crop, this paper

W. Wang (✉) · Z. Liu
Xiaolangdi Multipurpose Dam Project Management Center, Ministry of Water Resources, Zhengzhou 450000, China
e-mail: 232771432@qq.com

H. Xu (ed.), *Proceedings of the 5th International Symposium on Water Resource and Environmental Management*, Environmental Science and Engineering,
https://doi.org/10.1007/978-3-031-31289-2_12

constructs a GIS-based water demand forecasting system for the Yellow River diversion irrigation area. Through embedding GIS component library and tool library into the GIS development platform, using the ArcGis software and adopting a spatial and attribute information management method, the forecasting, processing, computing, analyzing and other functions of the irrigation area management are achieved.

12.2 Characteristics of GIS-Diversion Yellow River Irrigation Area Water Demand Forecasting System

In the process of using the GIS system to carry out water conveying in the Yellow River irrigation area, the specific application features include following aspects: (1) In the calculation and division of the water consumption units, it needs to perform operational division of the soil morphology, depth of groundwater, crop variety within the area, and project distribution characteristics in the water consumption area. During this process, GIS technology will divide farmland with same conditions into a same small system unit, at the same time, according to actual local conditions, the GIS system can overlay the relevant grid data and unit data, so as to calculate the water consumption amount of each area; (2) Meanwhile, for the prediction of irrigation area water consumption amount, during the application process, GIS system also can adopt local soil moisture modeling method to forecast the local soil moisture according to the change of time. In this function, the system can ensure that, when the soil water content reaches the minimum required value, the system can carry out irrigation work according to its actual situation. By adopting this calculation method, GIS system can accurately calculate the irrigation amount required by the crops within the irrigation area, the actual water requirement is determined by predictions of actual irrigation conditions and area and other parameters, so as to ensure normal operation of the system (Yan 2013).

12.3 Technology of GIS-Diversion Yellow River Irrigation Area Water Demand Forecasting System

12.3.1 System Operation Model

In order to ensure that the irrigation level in the Yellow River irrigation area is continuously improved under the guidance of the GIS system, it is necessary to ensure that the system operation model meets the requirements before applying the system. In this process, the application of GIS technology is mainly reflected in the following aspects: (1) GIS technology can make in-depth observation of the soil moisture content of local crops, especially identify the cultivated area of wheat, an important crop in the Yellow River irrigation area, and model to determine its moisture content index. In this process, GIS technology will analyze the soil moisture at different depths in different regions and use the method of remote sensing satellite imaging to determine; (2) Obtain the grid data of the soil moisture content in the current area of the irrigation area through remote sensing technology, these data represent an area that makes up the irrigation area, since the resolution of current remote sensing technology can reach within 1 m, in this process, we can set that the moisture content is the same, and then analyze its characteristics. That is, the soil moisture content in the target area is obtained through the grid data obtained through remote sensing technology; (3) During the process of constructing the operation model, the relevant soil type, groundwater depth, crop variety, and distribution project need to be calculated in detail and processed according to division, in this process, the actual local conditions need to be considered in detail and the actual planting scope need to be determined in order to realize the calculation of the local soil moisture content, so as to ensure timely irrigation when the local soil moisture content declines (Peng 2013). The relevant monitoring formulas are as follows (Tables 12.1 and 12.2):

(1) Prediction of soil moisture regression process

$$\theta_i = \theta_{i-1} - \frac{(ET_{i-1} - R_{i-1} - G_{i-1} - \Delta W + F_{i-1})}{\gamma H} \tag{12.1}$$

(2) Crop second-time irrigation water demand

$$I_i = (\theta_f - \theta_k) \times \gamma \times H_i \times \frac{2000}{3} \tag{12.2}$$

Table 12.1 Suitable soil moisture content for each growth stage of staple crops

Winter wheat	Growth stages						
Corresponding time period	Early Oct. to mid Dec.	Mid Dec. to late Feb.	Late Feb. to early Mar.	Early Mar. to early Apr.	Early Apr. to early May	Early May to late May	Late May to mid Jun.
Moisture content (relative) (%)	60–70	55–80	60–80	65–80	65–80	65–80	50–70
Summer corn	Growth stages						
Corresponding time period	Early Jun. to mid Jul.	Mid Jul. to early Aug.	Early Aug. to late Aug.	Late Aug. to early Sep.	Early Sep. to late Sep.		
Moisture content (relative) (%)	55–70	65–80	70–90	65–80	60–80		

Table 12.2 Irrigation system of irrigation areas

Irrigation start and stop time	Irrigation days	Irrigation quota (m^3/mu)	Irrigation quota (m^3/mu)
Mid Jun.	10	100	480
Late Jun.	10	60	
Early Jul.	10	60	
Mid Jul.	10	60	
Late Jul.	10	60	
Early Aug.	8	40	
Mid Aug.	8	50	
Early Sep.	8	50	

12.3.2 System Operation Function

After investigating the demand for irrigation water in the Yellow River irrigation area, the system operation function of GIS technology should be clearly integrated with local data and systems, so that the system can achieve the optimal combination of data and system groups in the design. In this process, the local space management data will be automatically incorporated into the scope of the observation system. The design method includes the following aspects:

(1) In the design of the GIS water demand forecasting system for the Yellow River irrigation area, the system can use local spatial and attribute information management methods to realize the functions of management forecasting, processing, calculating and analyzing, etc. This will make the entire water consumption system to be vividly and intuitively distributed, thus fulfilling the purpose of meeting local irrigation needs; (2) In the design process, designers need to physically go deep into the local area to establish input and preservation of spatial data in the irrigation area, which is not only conductive to implementation of local data standardization (Hou 2012), but also can realize spatial attribute query and irrigation of local area and output in the form of tables, so as to meet local water demand forecasting; (3) Yellow River water diversion project system has high system and organizational requirements. The GIS technology, as a computer system which uses digital form to collect, process, storage and analyze spatial data, is consisted of four parts: data, hardware, software, and application. To ensure its organization, the system should meet different conditions such as sharing, stability, security and scalability; (4) In order to ensure that the GIS is highly analytical in the processing process, a system flowchart can be set in it. The flow chart can classify the logical process of each processing step that occurs in the computer transactions from the start of data input until the output is obtained, and it can perform functional decomposition of the entire system from top to bottom based on the system requirements resulting from functional analysis to hierarchically determine the structure of the application program (Javier et al. 2008).

12.3.3 System Operation Structure

The GIS system can be divided into applied GIS system (mainly used for the processing of a certain specialty, field or work, including thematic geographic information system and regional integrated geographic information system) and tool GIS system (mainly has a variety of GIS application software packages, such as digital graphic images, storage management, query and search, analysis and operation, and output) according to actual needs, both kind have the following features in terms of functionality: (1) Applied GIS system is a spatial information system that has been widely implemented and has strong application. At present, in the GIS system, the development of applied GIS system has become increasingly important (Havaš and Jagić 2012). The system can accurately predict the digital raw information implemented in the irrigation area and carry out corresponding management work. Because the key issue in the entire R&D process is to target the application and efficiently develop a Yellow River water diversion irrigation water demand system that meets the demand and has a convenient and rich interface, therefore, this paper mainly uses the applied GIS system; (2) Tool GIS system has problems such as many limitations and its independent development is difficult, so it's not as convenient as the applied GIS system, in order to ensure that the actual application effect is efficient and simple, now a new combined GIS system has been developed which integrates all the advantages of the applied GIS system and the tool GIS system, this new system can fully utilize the visualization technology to achieve observation and irrigation operations in the measured area (Morović and Ivanov 2011).

12.3.4 System Operation Development

In the developing process of GIS Yellow River irrigation area water demand forecasting system, the designers need to use the relevant C language technology, the development environment is Microsoft.NET Framework, and an Access database is built based on the OFFICE components, the database contains the NET. Framework programs, the planned execution orders include following aspects:

(1) On the GIS development platform, designers need to use ArcGIS Engine to independently develop a set of embedded GIS component libraries and tool libraries that can be packaged, and develop and create a set of application data that is mainly used for exploration activities for its independence and completeness, the basic ArcGIS Engine generated by this data consists of two systems, the Developer Kit for building software and the second-release Runtime set by the completed application program. It supports development works under four environments: C++, COM, .NET, and Java. It can be implemented on Windows, UNIX and Linux platforms. Its independent embedded components can make it be independent from the ArcGIS Desktop platform, so that it can be directly applied to the toolkit development and application summary, so as to achieve development under different language environments; (2)

C language has a powerful function, while Visual Basic has a simple feature, in the design of the GIS system, we need to learn from the common advantages of both to organize the development so that both C++ and Java can be program-oriented languages; (3) In the design of the GIS system database, the processing of its organizational structure, including Oracle, Microsoft SQL Server and Access system, needs to be considered; (4) For the system's hardware environment, as GIS system has a high demand on its software data processing capabilities, its hardware must have advantages of high speed, good stability, sufficient capacity, and convenience.

12.3.5 System Operation Database

In order to ensure that the designed GIS system has a perfect water demand forecasting function, the designers also need to focus on the issues that appear in the design of spatial data and attribute data in the designed system. The relevant descriptions are as follows: (1) In the design of spatial data for GIS systems, designers need to pay attention to the use of the spatial database function carried by ArcGis to unify the management work of GIS system water demand management (Wang et al. 2018). In the process of work, the database need to start from the perspective of vector data and grid data, choose method to perform calculation on the system spatial data according to the element name, element type, and management attribute, so as to accurately obtain grid with groundwater depth data. (2) For the design of attribute data in the database, the designers need to perform their design work taking full advantage of the Access database provided by the Office system. In the design process, designers can take its advantages of simple operation and fast running speed, and use following information for calculation: regional sub-irrigation-area attribute table (including name of the region, regional rice area, regional wheat area, regional corn area, regional monthly well depth forecast value and regional irrigation canal length, etc.), regional seasonal average precipitation, regional irrigation water consumption factors (including irrigation area number, irrigation canal name, irrigation unit name, irrigation water consumption factor, and field water consumption coefficient and irrigation canal water consumption factor, etc.), geographic information in the irrigation area, planned irrigation depth, suitable water content, etc., finally, the corresponding soil moisture content is obtained (Münch and Conrad 2007).

12.3.6 System Operation Application

In order to ensure that the designed GIS system can actually realize the collection, sorting and configuration of the irrigation volume, irrigation area and crop structure of the irrigation area, the related working methods are as follows: (1) When the measured data is configured and measured, the designers need to use a variety of images in the same area to complete the coordinate matching of the relevant regions. The techniques used in the matching process include geometric correction, mapping transformation and unified processing of the three-azimuth scale. During the processing, system designers must select a ground control point to use the computer to control and correct its position (Ike 2018). The position setting needs to be configured with reference to the specific component standard in the image. The configuration work needs to use ArcGis software, using its configured map software to determine the middle coordinates to select the configurator ground control point, the selection needs to be determined according to the control point local distribution situation, control point identifiability, control point number and accuracy, and according to the vectorization standard as well; (2) At the same time, because the data obtained above is grid data, the designers need to re-arrange and combine them according to the grid form to determine the arrays and collections of different colors. This requires transforming the previous data into simpler vectorization data, so as to realize searching requirements of data and irrigation information. To choose a suitable forecasting model, we need to understand the law of the exploration data (Anonymous and Naghdi 2009).

12.4 Conclusion

In summary, this paper focuses on the operation characteristics of the Yellow River irrigation area water demand forecasting system, through researches, it's found that, during the actual application, GIS technology can fulfill the searching requirements of the data and irrigation information, make full use of visualization techniques to achieve observations and irrigation operations in the measured areas, so as to finally get a good effect of the corresponding soil moisture. It is hoped that the research in this paper can help the relevant personnel in the Yellow River irrigation area to use the GIS technology to carry out water demand forecasting for the Yellow River irrigation area, so as to contribute to the broader development prospect of local agriculture. After the completion of the GIS-based water demand forecasting system for the Yellow River diversion irrigation area, currently, there's only one pilot application that has been carried out in the People's Shengli Canal Irrigation Area. In the future, this system will need to run and use data from other irrigation area so that it can be further supplemented and improved, its functions need to be expanded and improved continuously so that it can become a mature and practical system.

References

Anonymous M, Naghdi M (2009) Research and markets: where in the city using GIS to manage government information (strategic focus) provides strategies to help vendors penetrate the government GIS industry. M2 Presswire 18–20

Havaš L, Jagić D (2012) Introducing GIS to HEP ODS d.o.o. elektra Zabok. Tech J 6:73–77

Hou JW (2012) Spatial optimization of water resources coupling between ACA and RS and GIS. Henan University
Ike CC (2018) Exponential Fourier integral transform method for stress analysis of boundary load on soil. Math Model Eng Probl 5:33–39
Javier N, Mónica T, Javier Z (2008) Validación del Cuestionario sobre Alteraciones de Reflujo (RDQ) y de la Escala de Impacto de la Enfermedad por Reflujo Gastroesofágico (GIS) para población Española. Gastroenterol Hepatol 32:22–25
Morović M, Ivanov YA (2011) Oil spill monitoring in the Croatian Adriatic waters: needs and possibilities. Acta Adriat 52:45–56
Münch Z, Conrad J (2007) Remote sensing and GIS based determination of groundwater dependent ecosystems in the Western Cape, South Africa. Hydrogeol J 15:19–28
Peng J (2013) Research on multi-objective dynamic allocation of water resources based on GIS. Tianjin University
Wang JT, Jing M, Cheng XG, Zhang X (2015) Research on water demand forecasting system for Yellow River irrigation district based on GIS. Water Sav Irrig 75:66–68
Wang XR, Ren GL, Zhang JX (2018) Numerical simulation and optimization analysis of thermal balance of heavy oil box-type substation louver arrangement. Math Model Eng Probl 5:21–26
Yan MY (2013) Research on GIS-based irrigation area water resources management information system. Zhengzhou University

Chapter 13
Biodiversity Associated to Floating Wetland Islands

C. S. C. Calheiros, M. Ilarri, and S. I. A. Pereira

Abstract Floating wetland islands (FWI) are nature-based solutions that can be applied in different water bodies, such lakes, ponds and rivers, considering a wide range of purposes. They have been considered for eutrophication abatement, wastewater treatment and ecosystem rehabilitation. They are of great value since their efficiency relies on bioremediation processes and are thus very versatile in terms of water depuration but also on biodiversity promotion. These systems comprise a floating platform, selected plants and an anchoring system. Depending on their configuration, they will attract and harbor different organisms such macrofauna and microorganisms, that will use the platform as shelter, habitat, stepping stone, nursery, food source among other benefits. The plant species are of particular interest since they are the living interface between the platform and the water body and they must be selected according to criteria that will enable their successful establishment having in consideration the biotic factors. The present paper intends to give an overview of the role of FWI towards their associated biodiversity and what has been identified in literature as the main groups that can be found and related to operational conditions. This approach can support future decisions concerning the FWI implementation conditions and components towards biodiversity enhancement.

Keywords Microbial communities · Macrophytes · Macrofauna · Macroinvertebrates · Nature-based solutions · Plant diversity

---

C. S. C. Calheiros (✉) · M. Ilarri
Interdisciplinary Centre of Marine and Environmental Research (CIIMAR/CIMAR), University of Porto, Novo Edifício do Terminal de Cruzeiros do Porto de Leixões, Avenida General Norton de Matos, S/N, 4450-208 Matosinhos, Portugal
e-mail: ccalheiros@ciimar.up.pt

S. I. A. Pereira
Universidade Católica Portuguesa, CBQF—Centro de Biotecnologia e Química Fina—Laboratório Associado, Escola Superior de Biotecnologia, Rua Diogo Botelho 1327, 4169-005 Porto, Portugal

H. Xu (ed.), *Proceedings of the 5th International Symposium on Water Resource and Environmental Management*, Environmental Science and Engineering,
https://doi.org/10.1007/978-3-031-31289-2_13

13.1 Introduction

The conservation and sustainable use of the marine and terrestrial environment is supported by the primary biodiversity Sustainable Development Goals (SDGs) 14 and 15 although, the 17 SDGs ultimately depend on healthy ecosystems and biodiversity. Concerning nature-based solutions (NBS), biodiversity has firstly a functional role, which underpins ecosystem functions and processes thus delivering ecosystem services and supporting conservation objectives (WWAP (United Nations World Water Assessment Programme)/UN-Water 2018). Nature-based solutions "are inspired and supported by nature and use, or mimic, natural processes to contribute to the improved management of water. A NBS can involve conserving or rehabilitating natural ecosystems and/or the enhancement or creation of natural processes in modified or artificial ecosystems" (WWAP (United Nations World Water Assessment Programme)/UN-Water 2018, p. 2).

The use of constructed wetlands as NBS for bioremediation has been widely used, considering wastewater treatment, management and ecosystems rehabilitation (Calheiros et al. 2020b). Although, when the target is a water body, such as a river, lake, pond, the use of floating wetland islands (FWI) is usually the most adequate approach. There are several reviews that detail the systems' applications and configurations (e.g., Arivukkarasu and Sathyanathan, 2023; Colares et al. 2020; Pavlineri et al. 2017). Floating wetlands islands are also known as treatment floating wetlands, floating islands or vegetated floating mats, and they deliver a wide range of ecosystem services, having the advantage that no additional surface land is required and can operate in situ. These systems comprise a floating platform, plants species and an anchoring system. In relation to the floating platform, in order to ensure the buoyancy, it can be made of different materials such polyethylene, polypropylene, polyurethane or polyvinyl alcohol foam (Karstens et al. 2021) or cork (Calheiros et al. 2020a). The plant species considered may vary widely depending on the environment (e.g., saline, freshwater) that the FWI will be implemented, being important to select them based on established criteria (Arivukkarasu and Sathyanathan, 2023; Calheiros et al. 2022; Wang and Sample 2014). The promotion of the biodiversity is one of the ecosystem services delivered by these systems and can be enhanced by adding other features to the platform in order support for example nursery purposes, habitat, shelter, etc. A huge interest is rising concerning the species that these systems can harbor at the macro and micro level.

The aim of this paper is to deliver an overview of the FWI role towards biodiverse promotion. In that sense it will be explored their associated biodiversity and what has been identified in literature as the main groups that can be found and the related operational conditions. This approach is of outmost importance to support future decisions concerning the FWI implementation conditions and components towards biodiversity enhancement.

13.2 Biodiversity Behind Floating Wetland Islands

In addition to the FWI's main purpose of treating water, these systems can also deliver numerous ecosystem services, including the improvement of biodiversity. The wide range of plant species used to vegetate the platforms will boost the biodiversity of the surrounding ecosystem. Starting with the increment of the diversity of plant species in the FWI itself. A study performed by Wang et al. (2015) showed that plant diversity increased from three (initially planted) to seven species after 1 year of the FWI implementation. The biofilm attached to the surface of roots and rhizomes and to the floating mat also comprises diverse bacterial, fungi, and algae communities. Moreover, FWI contribute to increase the diversity of several macroinvertebrates communities, namely insects, snails, and molluscs, while they create habitats for several animals, namely fishes, amphibians, and birds (Karstens et al. 2021).

13.2.1 Plant Selection and Establishment

Floating wetland islands consist in a buoyant structure floating on the surface of water that supports emergent vegetation. The roots of plants are completely emerged in the water, while the upper parts remain and grow above the water level (Arivukkarasu and Sathyanathan, 2023; Pavlineri et al. 2017; Benvenuti et al. 2018). The main natural components of FWI are the macrophytes which have extensive root systems enabling the absorption of nutrients and contaminants from water (Arivukkarasu and Sathyanathan, 2023). Macrophytes can be classified according to the water level as emergent, floating, and submerged (Thampatti et al. 2020). A plethora of plant species have been used for water treatment by FWI, however the selection of macrophytes is a crucial step and should rely on several criteria as they will impact the performance of the system, as well as the surrounding ecosystem.

Wang and Sample (2014) defined some criteria to select plants for FWI implementation:

- Native and non-invasive plants species
- Terrestrial and perennial plant species
- Emergent wetland plants able to adapt and thrive in hydroponic conditions
- Plants with aerenchyma for higher oxygen diffusion from aerial parts to belowground organs (roots and rhizomes)
- Plant species with aesthetic attributes
- Plants with high nutrient uptake capacity

In addition, Arivukkarasu and Sathyanathan (2023) reported that the selection of macrophytes should be based on following criteria:

- Saplings should be readily available
- Good adaptation to climatic conditions throughout the year
- Ease vegetative reproduction

- Good resistance to insects and plant diseases
- High growth rate and fast biomass production throughout the year
- High root development and anchoring well in the growth medium
- Provision of oxygen to the belowground organs even under anoxic conditions

Further on, Calheiros et al. (2020a) used the following criteria for plant selection: (1) native to the country; (2) potential to survive in hydroponic mode with tolerance to variable salinity levels; (3) presence in the region of the experimental site; (4) perennial species and (5) possibility to form a polyculture. Shahid et al. (2020) highlighted that the tolerance of plants to salinity, metals, organic compounds should also be considered.

The species mostly used in artificially created FWI belonged to genera *Canna, Carea, Typha, Phragmites, Juncus,* and *Cyperus* (Headley and Tanner 2012; Rehman et al. 2019). According to the meta-analysis performed by Pavlineri et al. (2017), *Canna* is a widely adapted plant species that is commonly used in FWI systems operating in several locations and climate zones in the world. Similarly, several species of *Typha* and *Cyperus* genera have also been broadly tested in FWI. For instance, the macrophyte *Typha domingensis* Pers. was successfully used in an FWI to treat raw sewage in a municipal sewage treatment plant (Benvenuti et al. 2018). However, many other plant species can be used, including terrestrial plant species such as *Brachiaria mutica* and *Salix babylonica.* These species are a good solution to increase the rate of removal of nutrients and contaminants from water as they possess extensive root systems (Ijaz et al. 2015; Zhu et al. 2011).

Queiroz et al. (2020) evaluated the use of the macrophytes *Eichhornia crassipes, Eichhornia paniculate, Polygonum ferrugineum* and *Borreria scabiosoides* for the treatment of dairy wastewater using FWI. The aquatic macrophytes *E. crassipes* and *E. paniculate* were the species that better performed showing higher organic load removal efficiency and biomass growth. Calheiros et al. (2020a) reported the successfully installation of an in-situ FWI in a seawater port marina, comprising a polyculture of species able to withstand high salt levels—*Sarcocornia perennis, Juncus maritimus, Phragmites australis, Halimione portulacoides, Spartina maritima* and *Limonium vulgare.*

Ornamental plants can also be used in FWI contributing to improve the aesthetic landscape (Barco and Borin 2020; Zanin et al. 2018), which could be of particular interest in urban areas. Moreover, ornamental plants will attract further wildlife, including insects pollinators. These plants must thrive and grow well in aquatic environment while showing ornamental features. Several strategies can be applied when using ornamental plants in FWI, namely the use of plants with a colorful flowering, mixtures of species with different colored flowers, and species with aesthetic foliage characteristics (Barco and Borin 2020). The suitability of 11 wetland species with ornamental features: *Canna indica* L., *Pontederia cordata* L., *Thalia dealbata* Fraser ex Roscoe, *Acorus calamus* L., *Juncus effusus* L., *Iris laevigata* L., *Mentha aquatica* L., *Oenanthe* javanica (Blume) DC., *Caltha palustris* L., *Sparganium erectum* L. and *Zantedeschia aetiopica* (L.) Srengel was evaluated. Results showed that *C. indica, P. cordata* and *T. dealbata* were the species that better performed regarding the vigor

Fig. 13.1 Examples of ornamental plant species used in floating wetland islands. **a** *Canna indica*, **b** *Canna flaccida*, **c** *Zantedeschia aethiopica*

and colonization of the floating maps, survival, and nitrogen uptake (Barco and Borin 2020) (Fig. 13.1). According to Liu et al. (2016) the ornamental emergent macrophytes, *Lythrum salicaria* Linn. and *Iris wilsonii*, were efficient in removing total nitrogen and phosphorous from the wastewater. Urakawa et al. (2017) also used *Canna flaccida* and *J. effusus* roots in an FWI established in a manmade stormwater pond.

13.2.2 Microorganisms Associated to Floating Wetland Islands

The roots of plants growing in the FWI, and the mat platform harbor a variety of microorganisms (Lubnow 2014; Lucke et al. 2019), which may be involved in the removal of total nitrogen and NH_4-N from water by several processes, such as nitrification, denitrification, and anaerobic ammonium oxidation (Tao and Wang 2009). Indeed, according to Sun et al. (2009), the removal efficiency of nitrogen was foster by the addition of immobilized denitrifying bacteria to the floating bed. In addition, microorganisms also play an important role on the degradation of pollutants, including metals and organic compounds (Liu et al. 2019), especially whether associated to the roots of plants in FWI (Rehman et al. 2019; Li et al. 2012; Demarco et al. 2023). Recent studies reported the inoculation of specific microorganisms in FWI systems as a successful strategy to increase the efficiency of pollutant removal from water (Shahid et al. 2020). For instance, Nawaz et al. (2020) demonstrated that the bacterial inoculation of a FWI composed of *P. australis* increased the removal of dyes from textile wastewater. Likewise, the inoculation of two aquatic plants with bacterial strains showing pollutant-degrading ability and growth-promoting traits fostered textile dye degradation (Tusief et al. 2022).

Fig. 13.2 **a** Floating wetland island in a freshwater pond, **b** biofilm sampling for microbial communities analysis

Despite the importance of microorganisms in nutrient and/or pollutant removal, little is known about the microbial communities associated to FWI plant roots (Urakawa et al. 2017). Preliminary studies are being carried out by Calheiros et al. (2022) that is studying the integration of an FWI pilot in a freshwater pond assessing its biodiversity (Fig. 13.2), namely the culturable bacterial communities associated to the floating platform biofilm and to the plant rhizosphere.

Urakawa et al. (2017) evaluated the microbial communities colonizing *C. flaccida* and *J. effusus* roots, as well as the biofilms of plant pots and floating mat foam, in an FWI implemented in a manmade stormwater pond. Results showed that FWI's plant root microbiomes were dominated by *Anabaena*, *Rhizobium*, and *Rhodobacter* genera. In addition, microbial communities of the FWI plant roots were quite different from the those found in the biofilm, being more related to the communities thriving in the surrounding water samples. Similar results were obtained by Calheiros et al. (2020a) in an FWI installed in a seawater port marine. The structure of the microbial communities retrieved from biofilm adhering the planting media and the submerged platform was distinct from the communities associated to the surrounding water.

13.2.3 Aquatic and Terrestrial Associated Fauna

Floating wetlands islands have real potential to provide habitats for aquatic and terrestrial fauna. Aquatic and terrestrial species usually use the available area as shelter and refuge to escape predator pressure (Karstens et al. 2021). Floating wetlands islands are also used by the fauna as a food source. The roots of plants in FWI are usually able to retain particulate matter that serve as food source for the macroinvertebrates (Prashant and Billore 2020). Aquatic and terrestrial associated species mainly include, annelids (mainly oligochaete worms and leeches), molluscs (mainly aquatic gastropods and freshwater mussels), arachnids (mainly spiders and water mites), crustaceans (mainly shrimps and crabs), insects (mainly dragonflies, bugs, water scorpions, water boatmen, beetles, caddisflies, midges), amphibians (American bull-frogs, gray treefrogs, crawfish frogs, southern leopard frogs), reptiles (Common snapping turtles, eastern painted turtles, red-eared slider turtles, northern water snakes), fishes (mainly eels, cyrprinids, gobies, rosy bitterlings, topmouth gudgeons, black basses) and birds (mainly mallards, red-winged blackbirds, twice gulls, grey herons, Canada gooses, sparrows) (Table 13.1).

Floating wetland islands can provide critical habitat also for many migratory fishes and invertebrate species important to commercial and recreational fisheries (Deegan et al. 2002; Moody et al. 2013). Indeed, Huang et al. (2017) reported that stereoscopic artificial floating wetlands vegetated with *P. australis* proved to be efficient fish aggregation devices in estuarine and coastal habitats. Recent studies have also observed that large numbers of juvenile European eel (*Anguilla anguilla*) are associated with the dense root structures of FWI plants (Cicero-Fernandez et al. 2022; Karstens et al. 2022). In this case, the FWI serve as a refuge to avoid predators; normally, eels seek out dark, structurally rich areas during the day (Karstens et al. 2021). This is of particular importance as the population of the European eel has declined in several places in recent years, and the species is nowadays considered critically endangered (Pike et al. 2020).

According to Kartens et al. (2022), the FWI can serve as a steppingstone to improve habitat connectivity of several species, such as the diadromous fish species. There are several bird species that are attracted to FWI and use the structure as an enhancement of available habitat (Yeh et al. 2015). Shealer et al. (2006) observed that during periods of flooding birds preferred FWI over natural sites. They have also observed that bird eggs that were laid in FWI were significantly larger than those laid at natural sites, suggesting that FWI were occupied by better quality birds. According to Hancock (2000), the availability of FWI have probably increased the Black-throated Diver (*Gavia artica*) population in Scotland by an average of 44% (from 7 to 130%, varies by site). The FWI appears to have been used by birds as an additional benefit contributing to landscape aesthetics and also habitat enhancement.

Overall, FWI seem to improve ecosystem conditions and lead to more diverse fauna (invertebrates, fishes, amphibians, reptiles, and birds) communities. In particular, FWI provide shelter when they provide more interstitial spaces favorable to the colonization of macroinvertebrates; heterogeneous habitats that contributes to attract diverse and abundant fauna; stable and constantly accessible habitats for fauna regardless of the water level, as they are floating structures; and they also serve as a relevant food source for some species.

13.3 Conclusion

Floating wetland islands are very interesting NBS to consider when a water body needs of intervention being for water management and treatment or ecosystem rehabilitation or aesthetic integration. They are very versatile and adaptative in terms of their configuration and components. Enhancement of biodiversity is an asset

Table 13.1 Main species and groups observed in association with the floating wetlands islands (FWI)

Species	Main use of the floating wetland	Period of study	System	Reference
Annelida (2 species), Mollusca (2 species), Crustacean (*Macrobrachium niponense*), Arachnida (*Aranae* sp.), Insects (Terrestrial, 4 groups; Aquatic, 4 groups), Fishes (6 species)	Favorable habitat, safe nesting habitat, area for feeding	3 years	Shoreline of a lake	Nakamura et al. (1997)
Annelida (Freshwater leech), Mollusca (*Physella acuta*), Arachnida (*Dolomedes triton*), Insects (Aquatic, 4 species; Terrestrial, 5 species), Fishes (*Cyprinus carpio*), Reptiles (4 species), Birds (5 species)	Foraging, breeding, nursing, resting	5 months	Pond	Wang et al. (2015)
Crustaceans (Crabs, 8 species; Shrimps, 5 species), Fishes (13 species)	Used to compensate habitat degradation, improve fish and crustacean recruitment and increase refuge capacity	Few days during 4 months	Reservoir	Huang et al. (2017)
Mollusca (Gastropods, 2 species), Arachnida (2 species), Insects (3 species), Reptiles (4 species), Birds (3 species)	Habitat creation	2 years	Ponds	Strosnider et al. (2017)

(continued)

Table 13.1 (continued)

Species	Main use of the floating wetland	Period of study	System	Reference
Annelida (Polychaeta), Crustaceans (*Chthamalus* sp.; *Palaemon* sp.), Bivalvia (*Mytilus* sp.), Fishes (*Mugil* sp.)	Feeding	16 months	Next to a pier of a Marina	Calheiros et al. (2020a)
Mollusca (7 species), Insects (11 species)	Shelter and food availability	Not available	River	Prashant and Billore (2020)
Insects (*Nepa* sp.), Crustaceans (2 species), Mollusca (*Gastropoda indet*.), Fishes (*Anguilla anguilla*), Amphibians (*Rana esculenta*), Reptiles (*Natrix natrix*), Birds (5 species)	Foraging, shelter and refuge	4 months	Shallow lagoon	Karstens et al. (2021)
Crustaceans (2 species), Fishes (*Anguilla anguilla*)	Functional habitat, nursery and refuge	2 years	Floating dock in the tidal lagoon	Cicero-Fernandez et al. (2022)
Cnidaria (*Hydra* sp.), Annelida (Oligochaetes), Mollusca (3 species), Arachnida (*Hydracarina* sp.), Insects (5 species)	Diverse habitats for the different stages of macroinvertebrates	Seasonal (Autumn 2018, Winter 2019, Spring 2019, Summer 2019)	Bays in the downstream part of a reservoir	Salmon et al. (2022)

associated to these systems above and below water surface. Biodiversity can thus be promoted when choosing the adequate materials, especially the plant species, that are the most visible component being in the living interface. A wide range of biodiversity has already been identified to be associated to FWI, such, macroinvertebrates, vertebrates and microorganisms. Nevertheless, there is still the need for further research on the ecosystem's services accounting related to biodiversity, especially species association to different environments (marine, freshwater, wastewater) and their response to different biotic and abiotic profiles.

Acknowledgements This research was supported by National Funds from FCT-Fundação para Ciência e Tecnologia within the scope of projects UIDB/04423/2020, UIDP/04423/2020, and UIDB/50016/2020 projects.

References

Arivukkarasu D, Sathyanathan R (2023) Floating wetland treatment an ecological approach for the treatment of water and wastewater—a review. Mater Today: Proc. 77: 176-181

Barco A, Borin M (2020) Ornamental plants for floating treatment wetlands: preliminary results. Ital J Agron 15:109–120

Benvenuti T, Hamerski F, Giacobbo A, Bernardes AM, Zoppas-Ferreira J, Rodrigues MAS (2018) Constructed floating wetland for the treatment of domestic sewage: a real-scale study. J Environ Chem Eng 6:5706–5711

Calheiros CSC, Carecho J, Tomasino MP, Almeida CMR, Mucha AP (2020a) Floating wetland islands implementation and biodiversity assessment in a Port Marina. Water 12:3273

Calheiros CSC, Pereira SIA, Castro PML (2020b) Constructed wetlands as nature-based solutions. In: Jespersen AN (ed) An introduction to constructed wetlands, chap 3. Nova Science Publishers, Inc

Calheiros CSC, Ilarri M, Castro PML, Pereira SIA (2022) Floating wetland islands to support biodiversity promotion in a freshwater pond. In: Abstract proceedings of WREM 2022 conference: 2022, the 5th international symposium on water resource and environmental management, 9–10 Dec 2022, Webinar

Cicero-Fernandez D, Expósito-Camargo JA, Peña-Fernandez M (2022) Efficacy of Juncus maritimus floating treatment saltmarsh as anti-contamination barrier for saltwater aquaculture pollution control. Water Sci Technol 85(10):2811

Colares GS, Dell'Osbel N, Wiesel PG, Oliveira GA, Lemos P, da Silva HZ, Lutterbeck FP, Kist CA, Colares LT, Machado ÊL (2020) Floating treatment wetlands: a review and bibliometric analysis. Sci Total Environ 714:136776

Deegan LA, Hughes JE, Rountree RA (2002) Salt marsh ecosystem support of marine transient species. In: Concepts and controversies in tidal marsh ecology. Springer, Dordrecht, pp 333–365

Demarco CF, Carlos FS, Pieniz S, Morselli LBGA, Andreazza R (2023) Bioremediation of aquatic environments contaminated with heavy metals: a review of mechanisms, solutions and perspectives. Sustainability 15:1411

Hancock M (2000) Artificial floating islands for nesting Black-throated Divers Gavia arctica in Scotland: construction, use and effect on breeding success. Bird Study 47(2):165–175

Headley TR, Tanner CC (2012) Constructed wetlands with floating emergent macrophytes: an innovative stormwater treatment technology. Crit Rev Environ Sci Technol 42:2261–2310

Huang X, Zhao F, Song C, Gao Y, Geng Z, Zhuang P (2017) Effect of stereoscopic artificial floating wetlands on nekton abundance and biomass in the Yangtze Estuary. Chemosphere 183:510–518

Ijaz A, Shabir G, Khan QM, Afzal M (2015) Enhanced remediation of sewage effluent by endophyte-assisted floating treatment wetlands. Ecol Eng 84:58–66

Karstens S, Langer M, Nyunoya H, Čaraitė I, Stybel N, Razinkovas-Baziukas A, Bochert R (2021) Constructed floating wetlands made of natural materials as habitats in eutrophicated coastal lagoons in the Southern Baltic Sea. J Coast Conserv 25(4):1–14

Karstens S, Dorow M, Bochert R, Stybel N, Schernewski G, Mühl M (2022) Stepping stones along urban coastlines-improving habitat connectivity for aquatic fauna with constructed floating wetlands. Wetlands 42:76

Li H, Hao H, Yang X, Xiang L, Zhao F, Jiang H, He Z (2012) Purification of refinery wastewater by different perennial grasses growing in a floating bed. J Plant Nutr 35(1):93–110

Liu JL, Liu JK, Anderson JT, Zhang R, Zhang ZM (2016) Potential of aquatic macrophytes and artificial floating island for removing contaminants. Plant Biosyst 150:702–709

Liu T, Xu S, Lu S, Qin P, Bi B, Ding H, Liu Y, Guo X, Liu X (2019) A review on removal of organophosphorus pesticides in constructed wetland: performance, mechanism and influencing factors. Sci Total Environ 651:2247–2268

Lubnow FS (2014) Using floating wetland islands to reduce nutrient concentrations in lake ecosystems. Natl Wetl Newsl 36:14–17

Lucke T, Walker C, Beecham S (2019) Experimental designs of field-based constructed floating wetland studies: a review. Sci Total Environ 660:199–208

Moody RM, Cebrian J, Heck KL Jr (2013) Interannual recruitment dynamics for resident and transient marsh species: evidence for a lack of impact by the Macondo oil spill. PLoS ONE 8(3):e58376

Nakamura K, Tsukidate M, Shimatani Y (1997) Characteristic of ecosystem of an artificial vegetated floating island. Trans Ecol Environ 16:171–181

Nawaz N, Ali S, Shabir G, Rizwan M, Shakoor MB, Shahid MJ, Afzal M, Arslan M, Hashem A, Abd_Allah EF, Alyemeni MN, Ahmad P (2020) Bacterial augmented floating treatment wetlands for efficient treatment of synthetic textile dye wastewater. Sustainability 12:3731

Pavlineri N, Skoulikidis NT, Tsihrintzis VA (2017) Constructed floating wetlands: a review of research, design, operation and management aspects, and data meta-analysis. Chem Eng J 308:1120–1132

Pike C, Crook V, Gollock M (2020) *Anguilla anguilla*. The IUCN Red List of Threatened Species 2020, e.T60344A152845178

Prashant, Billore SK (2020) Macroinvertebrates associated with artificial floating islands installed in River Kshipra for water quality improvement. Water Sci Technol 81(6):1242–1249

Queiroz RCS, Lôbo IP, Ribeiro VS, Rodrigues LB, Almeida Neto JA (2020) Assessment of autochthonous aquatic macrophytes with phytoremediation potential for dairy wastewater treatment in floating constructed wetlands. Int J Phytoremediation 22(5):518–528

Rehman K, Ijaz A, Arslan M, Afzal M (2019) Floating treatment wetlands as biological buoyant filters for wastewater reclamation. Int J Phytorem 21:1273–1289

Salmon Q, Colas F, Westrelin S, Dublon J, Baudoin JM (2022) Floating Littoral Zone (FLOLIZ): a solution to sustain macroinvertebrate communities in regulated lakes? Ecol Eng 176:106509

Shahid MJ, AL-Surhanee AA, Kouadri F, Ali S, Nawaz N, Afzal M, Rizwan M, Ali B, Soliman MH (2020) Role of microorganisms in the remediation of wastewater in floating treatment wetlands: a review. Sustainability 12:5559

Shealer DA, Buzzell JM, Heiar JP (2006) Effect of floating nest platforms on the breeding performance of Black Terns. J Field Ornithol 77(2):184–194

Strosnider WH, Schultz SE, Johnson Strosnider KA, Nairn RW (2017) Effects on the underlying water column by extensive floating treatment wetlands. J Environ Qual 46:201–209

Sun L, Liu Y, Jin H (2009) Nitrogen removal from polluted river by enhanced floating bed grown canna. Ecol Eng 35:135–140

Tao W, Wang J (2009) Effects of vegetation, limestone and aeration on nitritation, anammox and denitrification in wetland treatment systems. Ecol Eng 35:836–842

Thampatti KCM, Beena VI, Meera AV, Ajayan AS (2020) Phytoremediation of metals by aquatic macrophytes. In: Shmaefsky BR (ed) Phytoremediation, concepts and strategies in plant sciences. Springer, pp 153–204

Tusief Q, Mohsin M, Malik MH, Asghar HN, Sardar S (2022) Textile wastewater treatment through plant–microbe synergism in a floating treatment wetland system. AATCC J Res 9(4):182–193

Urakawa H, Dettmar DL, Thomas S (2017) The uniqueness and biogeochemical cycling of plant root microbial communities in a floating treatment wetland. Ecol Eng 108(Part B):573–580

Wang CY, Sample DJ (2014) Assessment of the nutrient removal effectiveness of floating treatment wetlands applied to urban retention ponds. J Environ Manage 137:23–35

Wang CY, Sample DJ, Day SD, Grizzard TJ (2015) Floating treatment wetland nutrient removal through vegetation harvest and observations from a field study. Ecol Eng 78:15–26

WWAP (United Nations World Water Assessment Programme)/UN-Water (2018) The United Nations world water development report 2018: nature-based solutions for water. UNESCO, Paris

Yeh N, Yeh P, Chang Y (2015) Artificial floating islands for environmental improvement. Renew Sustain Energy Rev 47:616–622

Zanin G, Bortolini L, Borin M (2018) Assessing stormwater nutrient and heavy metal plant uptake in an experimental bioretention pond. Land 7:1–16

Zhu L, Li Z, Ketola T (2011) Biomass accumulations and nutrient uptake of plants cultivated on artificial floating beds in China's rural area. Ecol Eng 37:1460–1466

Chapter 14
Dynamic Evolution of Landscape Pattern in Hanzhong Pingchuan Section of Hanjiang River in China

Xiaomei Kou, Qi Li, Hong Zhao, Yongxiang Cao, and Le Niu

Abstract Understanding the dynamic change of landscape pattern is crucial for improving the ecosystem services and regulating water resources in the basin. In this paper, satellite remote sensing TM image data in different periods since the late 1980s were collected, combined with 1:1000 topographic map and other ground observation records, and taking into account the impact of human factors on the Hanzhong Pingchuan section of the Hanjiang River (HPHR). ARCGIS and FRAGSTATS spatial analysis modules were used to calculate the regional landscape pattern index and analyze the landscape pattern changes at different levels. The results showed that the major types of LULC of the HPHR included water areas cultivated land and woodland in 2013. During the whole study period, the area of woodland, construction land, water area and other land types increased, while the area of farmland and grassland decreased. The loss of landscape connectivity is often a direct result of fragmentation caused by human activities. The construction of land patches is mainly concentrated in the flat areas on both sides of the main stream while there are few on both sides of the tributaries and the ecosystem service value of the whole region is not fully reflected. The government and manager should optimize the regional landscape pattern, build the ecological corridor, enhance the continuity between patches. At the same time, the weak ecological function on the ecological corridor should be identified as the ecological node to provide a reasonable landscape pattern for the regional planning of the HPHR.

Keywords Hanjiang River basin · Landscape pattern · Plaque continuity · Landscape uniformity

X. Kou (✉) · Q. Li · H. Zhao · Y. Cao · L. Niu
Power China Northwest Engineering Corporation Limited, Xi'an 710065, Shaanxi, China
e-mail: Kouxm0427@nwh.cn

X. Kou · Q. Li · Y. Cao
Shaanxi Union Research Center of University and Enterprise for River and Lake Ecosystems Protection and Restoration, Xi'an 710065, Shaanxi, China

H. Xu (ed.), *Proceedings of the 5th International Symposium on Water Resource and Environmental Management*, Environmental Science and Engineering,
https://doi.org/10.1007/978-3-031-31289-2_14

14.1 Introduction

Land use/cover changes have significantly changed ecosystem model and process, resulting in changes in landscape patterns (Shirmohammadi et al. 2020; Qin and Chen 2023). Analysis of landscape pattern change and LUCC are inseparable. Landscape pattern represents the landscape ecological condition and indicates the spatial characteristics of landscape composition and configuration. The landscape pattern changes affect a series of ecosystem services, such as soil and water conservation, biological diversity, carbon sequestration and biogeochemical cycle of terrestrial ecosystems (McGarigal et al. 2012). Therefore, the study of the spatiotemporal evolution characteristics of landscape pattern can provide important guidance for protecting regional ecological environment and determining whether land use is reasonable, and also plays an important role in formulating watershed landscape planning and carrying out watershed comprehensive management (Wang et al. 2022; Feng et al. 2018).

Over recent decades, with the development of economy, the change of land use has led to the imbalance of the proportion of landscape types and the change of spatial allocation (Sunsanee and Rajendra 2016; Song et al. 2020). Spatial analysis tools like GIS, remote sensing, land-use models and statistical methods have been used to intuitively quantify the spatial heterogeneity of landscape pattern (Odongo et al. 2019; Xia et al. 2021). Studies have shown that the change of landscape pattern is the direct response of land use change (Yang 2007; Wu 2005). LULC types with a high ecological value, such as forests, grasslands and water bodies, but they are frequently affected by human activities, resulting in different degrees of degradation of ecosystem services (Cao et al. 2022; Huang et al. 2019; Cai and Hu 2008).

The HPHR is located at the upstream of the main stream of the Hanjiang River, with a length of about 119.5 km and a drainage area of 14,200 km^2. The protection and construction of ecological environment in HPHR is the key to guarantee the water quality safety of water sources (Wang et al. 2006). With the rapid development of regional economy and the acceleration of urbanization process. Human activities have caused changes in the underlying surface of the basin, such as soil and water conservation measures, city building, land use changes and water conservancy project construction, etc. The LULC in the basin has a great impact on runoff, vegetation and ecological environment. Within this context, the water ecosystem in the HPHR is under a certain degree of stress (Wu et al. 2005). Under this context, it is necessary to analyze the LULC and landscape pattern change at different levels in the region. The main purpose of this study is to analyze the land-use change, landscape change and provide data and policy support for the regulation of the regional landscape pattern.

14.2 Materials and Methods

14.2.1 Study Area

HPHR is located in the hinterland of Hanzhong Basin at the upstream of the main stream of Hanjiang River, it starts from Wuhou Town of Mian County in the west and ending at Xiaoxiakou of Yangxian County in the east. There are 27 main tributaries along the Hanzhong Pingchuan section, such as Yan River, Bao River, Lianshui River, Lengshui River, Xushui River, Yishui River, etc. The HPHR have a subtropical humid monsoon climate zone in the north of central China. Affected by the terrain conditions, it has formed a unique climate feature of Qinba Mountains and the climate has the characteristics of warm and humid, no extreme heat in summer, no severe cold in winter, and hot rain in the same season. Yellow brown soil is the main zonal soil of HPHR, its surface soil contains more humus, has light texture and loose structure, mainly distributed among low-elevation hills and terraces with an altitude of 800–900 m. HPRP has a wide variety of plants and rich resources (Su et al. 2008; Kang et al. 2015).

14.2.2 Data Sources

Combined with the 1:1000 topographic map and other ground observation data, and taking into account the degree of human factors on the image of the HPHR, three periods (1987, 2000, 2013) of Landsat TM images (spatial resolution 30 m) were gained from the Chinese Academy of Sciences. The images with cloud cover of less than 5% and good imaging quality from June to October of selected year. A series of image preprocessing works are carried out for remote sensing images, including geometric registration, radiation processing and normalization, image mosaicing and clipping (Khalyani and Mayer 2013). By analyzing the characteristics of land use in the HPHR, land is classified into 6 types, including cultivated land (CL), woodland (WD), grassland, Construction land (Con L), River surface and inland tidal flat (RSITF), and other land (OL). Classification was done in ERDAS IMAGINE 8.7 and the overall classification accuracy was verified exceeding 87%.

14.2.3 Methods

Landscape transfer matrix. The Markov model has been widely used to predict LULC changes which can be used to predict the transfer source and destination of land use type (Chen et al. 2018). The formula is as follows:

$$P_{ij} = \begin{bmatrix} P_{11} \cdots P_{1n} \\ \vdots \\ P_{n1} \cdots P_{nm} \end{bmatrix} \tag{14.1}$$

where P_{ij} represents the conversion probability of each land use/cover type, n and m represents the total number of different land use types, i and j represent the land use/cover types at the beginning and end of different time periods during the study period.

Landscape metrics. The quantitative description of landscape pattern analysis is mainly through calculation and analysis of the landscape index, so as to complete the quantitative analysis of overall characteristics of the landscape spatial pattern in the study area, as well as quantitative analysis of different landscape types in the region (Chen et al. 2022; Corry 2005). Spatial analysis modules of ARCGIS and FRAGSTATS were used to calculate the landscape pattern index (Mcgarigal and Marks 1995). These indexes mainly include the number, area, shape, structure, diversity and other indicators of patches, and have a strong ability to describe the landscape pattern (Chen et al. 2002). The main indicators we analyzed are as follows:

(1) Number of patches (NP):

$$NP = N \tag{14.2}$$

where N is the total number of patches.

(2) Mean Patch Size (MPS):

$$MPS = \sum_{j=1}^{n} a_{ij}/n \tag{14.3}$$

where MPS is equal to the total landscape area divided by the total number of patches of each type.

(3) Boundary Density Index (ED):

$$ED = \sum P / \sum A \tag{14.4}$$

where P is the perimeter of plaque. The complexity of the landscape patch shape is reflected in the ED.

(4) Mean Patch Fractal Dimension (MPED):

$$MPED = \frac{1}{N} 2ln\frac{1}{4}P/ln(A) \tag{14.5}$$

where P is the perimeter of the plaque. The complexity of patch shape can be measured by its fractal dimension and can also be used to describe the geometric complexity of landscape patches.

(5) Patch Shape Index:

Applies to vector data if using the circle as a reference geometry:

$$MSI = \frac{P}{2\sqrt{\pi A}} \tag{14.6}$$

Applies to grid data if using the Square as a reference geometry:

$$MSI = \frac{P}{4\sqrt{A}} \tag{14.7}$$

where *MSI* reflecting the development degree of the patch. The increase of *MSI* indicates the decrease of human interference.

(6) Landscape Diversity Index:

$$H = -\sum_{k=1}^{n} P_k ln P_k \tag{14.8}$$

where P_k is the probability of patch type k appearing in the landscape. H can reflect the heterogeneity of the landscape.

(7) Landscape Evenness Index:

The expression of Shannon Evenness Index (SHEI) is:

$$E = \frac{H}{H_{max}} = \frac{-\sum_{k=1}^{n} P_k ln P_k}{ln(n)} \tag{14.9}$$

where H is the Shannon diversity index, and H_{max} is its maximum value. Obviously, when E tends to 1, the uniformity of landscape patch distribution tends to be the largest.

14.3 Results

14.3.1 Dynamic Evolution of LULC in HPRP

Figure 14.1 showed the land use in different periods in the HPHR, Table 14.1 and Fig. 14.2 were the results of the classification of land use type in the survey area. In general, the largest type of land use is the river surface and inland tidal flat (including river surface and inland tidal flat), followed by forest land and cultivated land. In different periods, the proportion of river surface and inland tidal flat area is about 30%, but the proportion showed a decreasing trend. The area of cultivated land was 49.28 km^2 in 1987, increased to 80.63 km^2 in 2000, decreased to 61.74 km^2 in 2013 and the overall trend of land reclamation was upward. The area of forest land accounted for 20.73% in 1987, and decreased sharply to 13.46% in 2000. However, the area of forest land increased in 2013, accounting for 18.05% and the area of forest land showed a decreasing trend. The grassland area has changed greatly, with 6.11 km^2 in 1987 and 34.86 km^2 in 2000, but decreased in 2013, with an area of 30.48 km^2, showing an upward trend. The area of construction land (including residential land and transportation land) was 49.84 km^2 in 1987, decreased to 23.41 km^2 in 2000, and increased to 39.67 km^2 in 2013. Other land (mainly bare land) continued to decrease, from 3.78 km^2 in 1987 to 0.16 km^2 in 2000, and to 0.13 km^2 in 2013.

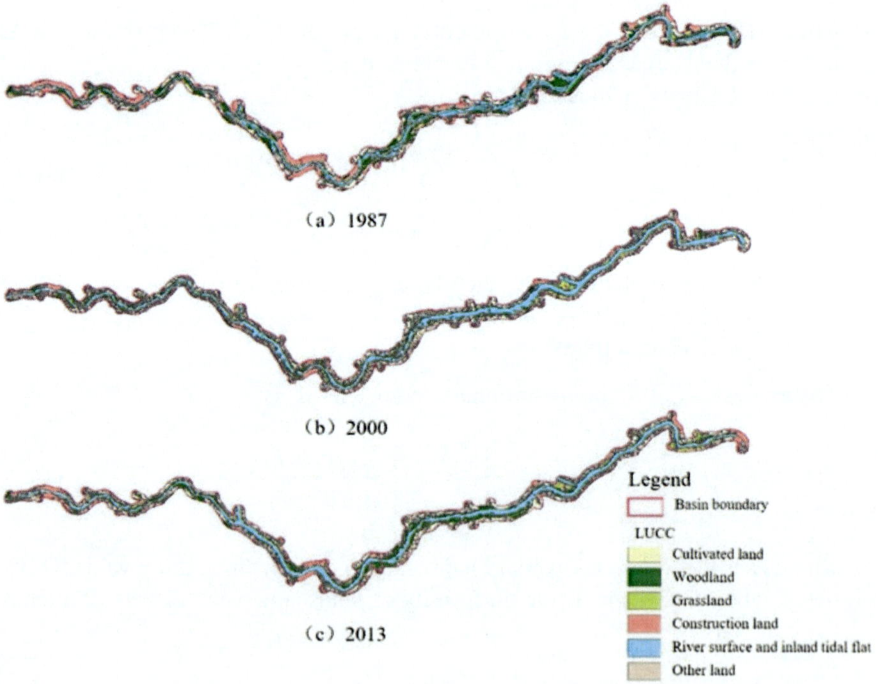

Fig. 14.1 Land use distribution in different periods in the HPHR

Table 14.1 Area of each land-use type in different periods

LULC	1987		2000		2013	
	Area (km^2)	Proportion (%)	Area (km^2)	Proportion (%)	Area (km^2)	Proportion (%)
CL	49.28	20.75	80.63	33.95	61.74	25.99
WL	49.24	20.73	31.96	13.46	42.86	18.05
GL	6.11	2.57	34.86	14.68	30.48	12.83
Con L	49.84	20.98	23.41	9.86	39.67	16.70
RSITF	79.26	33.37	66.49	27.99	62.63	26.37
OL	3.78	1.59	0.16	0.07	0.13	0.05

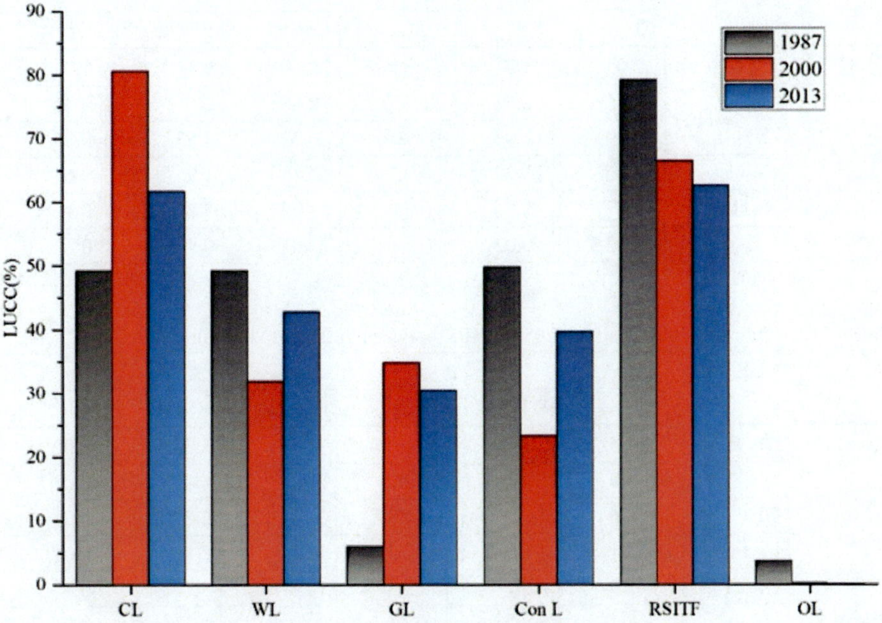

Fig. 14.2 Statistical map of land use in different periods in the HPHR

14.3.2 Analysis of Dynamic Change of LULC

The land-use change processes are presented with the transition matrix (see Tables 14.2, 14.3, and 14.4). There were significant changes in land cover and land use during the study period. In period from 1887 to 2000, cultivated land decreased by 31.36 km^2, forest land increased by 17.23 km^2, grassland decreased by 28.71 km^2, construction land increased by 26.35 km^2, water area and water conservancy facilities land decreased by 12.77 km^2, and other land increased by 3.72 km^2. The main reason for the land use change is that since 1987, the population of the HPHR has gradually become denser, and the construction land has increased. At the same time, the implementation of flood control and other projects in the Hanjiang River channel has reduced the area of farmland around the original river channel, which has been converted into forest land and grassland. The change of water area and water conservancy facilities land is mainly reflected by the image of 1987 in the year of high water and 2000 in the year of low water, resulting in drastic changes in the water area.

In period from 2000 to 2013, cultivated land increased by 18.88 km^2, forest land decreased by 10.91 km^2, grassland increased by 4.38 km^2, construction land decreased by 16.23 km^2, water area and water conservancy facilities land increased

Table 14.2 Transition matrix of LULC in PHPR between 1987 and 2000

Year	LULC	2000					
		CL	WL	GL	Con L	RSITF	OL
1987	CL	27.03	21.71	2.34	18.59	9.11	1.7
	WL	5.15	8.51	1.71	9.48	6.6	0.46
	GL	9.49	12.03	0.99	6.92	4.53	0.85
	Con L	3.86	3.22	0.35	13.44	1.9	0.6
	RSITF	3.57	3.66	0.71	1.29	57.1	0.16
	OL	0.02	0.02	0	0.01	0	0

Table 14.3 Transition matrix of LULC in PHPR between 2000 and 2013 (unit: km^2)

Year	LULC	2013					
		CL	WL	GL	Con L	RSITF	OL
2000	CL	31.75	6.1	14.48	3.91	5.38	0.02
	WL	16.49	11.68	5.87	2.34	6.45	0.01
	GL	12.06	4.91	8.88	2.32	2.27	0.01
	Con L	15.01	4.67	3.75	13.77	2.38	0.01
	RSITF	5.19	4.56	1.84	1.03	50.01	0
	OL	0.01	0	0.01	0	0	0.01

Table 14.4 Transition matrix of LULC in PHPR between 2000 and 2013

Year	LULC	2013					
		CL	WL	GL	Con L	RSITF	OL
1987	CL	31.71	12.25	1.34	10.24	5.16	0.91
	WL	3.33	16.11	1.95	7.02	13.12	1.3
	GL	5.32	11.46	1	8.7	3.6	0.34
	Con L	7.81	5.15	0.9	21.85	2.85	1.05
	RSITF	0.95	4.18	0.91	1.9	54.51	0.18
	OL	0.01	0.01	0	0	0	0

by 3.86 km^2, and other land increased by 0.02 km^2. The main reasons for land use change are rapid population growth, the increased demand for agricultural land, the conversion of large areas of forest land and construction land into cultivated land, the reduction of forest land and construction land and the increase of cultivated land. Remote sensing images show that the years of 2000 and 2013 are dry years, and the implementation of the comprehensive renovation of the Han River and other projects has also leading to the increase of water areas and water conservancy facilities.

In period from 1987 to 2013, cultivated land decreased by 12.48 km^2, forest land increased by 6.33 km^2, grassland decreased by 24.33 km^2, construction land increased by 10.11 km^2, water area and water conservancy facilities land increased by 16.62 km^2, and other land increased by 3.75 km^2. The change of land use during this period includes the conversion of cultivated land to construction land due to the increase of land demand, as well as the conversion of forest land and construction land to cultivated land. From the perspective of land use change, the implementation of the comprehensive regulation of the Han River and the flood control project of the river channel has resulted in the reduction of cultivated land and grassland, and the increase of forest land, water area, water conservancy facilities and construction land.

14.3.3 The Change Characteristics of Landscape Patches

At the level of patch type, the landscape pattern of 6 main land use types, including cultivated land, forest land, grassland, construction land, river surface and inland tidal flat and other land is analyzed. The result of landscape structure analysis shows that a decrease in landscape diversity, and an increase in dominance indices during the study period (Table 14.5).

The patch numbers of cultivated land increased significantly from 1987 to 2000, and decreased rapidly from 2000 to 2013. The main reason is that as population density in HPHR increases, the demand for agricultural land increased, and the area of cultivated land increased, so the patch numbers increased after 1987. The boundary density of cultivated land first increased and then decreased, the fractal dimension gradually increased and remained unchanged until 2013. The shape index first decreased and then increased, which indicates that the patch was gradually broken, the degree of interference was strengthened, and the patch geometry was more complex between 1987 and 2000. With the implementation of the Hanjiang Wetland Ecological environment Treatment Project, the fragmentation of cultivated land patches decreased and the geometric shape of the patch became simpler from 2000 to 2013. The number of forest patches increased significantly from 1987 to 2000, but decreased from 2000 to 2013. The boundary density and average patch fractal dimension also have a trend of increasing, which indicates that the form of forest land is more diversified. During the study period, the increase of road protection forest, farmland protection forest and observation platform green space in the study area made the spatial shape of the forest land more diversified and the degree

Table 14.5 Characteristics of landscape type patches in HPHR

LULC	Year	NP	MPS	ED	MPED	MSI
CL	1987	138	35.71	21.90	1.29	1.80
	2000	911	8.96	63.14	1.34	1.58
	2013	691	12.29	55.79	1.34	1.64
WL	1987	158	31.16	23.90	1.30	1.90
	2000	751	4.19	29.22	1.32	1.38
	2013	525	8.11	30.47	1.33	1.50
GL	1987	34	17.96	3.47	1.30	1.69
	2000	689	13.02	63.14	1.34	1.71
	2013	839	2.76	26.47	1.33	1.39
Con L	1987	205	24.31	23.15	1.30	1.72
	2000	212	4.44	7.60	1.32	1.35
	2013	672	8.39	42.11	1.33	1.54
RSITF	1987	162	48.93	39.37	1.31	1.98
	2000	156	16.08	16.88	1.34	1.70
	2013	111	27.38	16.80	1.33	1.76

of interference reduced. The number patches of grassland increased significantly and the patch density showed a trend of first significant increase and then decline from 1987 to 2000. The mean patch fractal dimension and shape index increased at first and then decreased, which indicates that the area of grassland increased, the number of patches increased and the patches gradually broken during this period. After 2000, the number of grassland patches increased but the dominant patches of grassland landscape decreased. The number patches of construction land increased rapidly during the study period, which shows that the construction land was gradually expanding. The boundary density and patch shape index decreased first and then increased. In particular, the change of boundary density is relatively significant, indicating that the patch shape of construction land has a trend of simple first and then complex. It can be explained by the fact that construction land, scattered rural homestead appeared, and the original scattered urban land was merged with the increased urban land during the study period. The number of landscape patches in the water area decreased between 2000 and 2013, mainly manifested by the disappearance of lower grade river channels. Based on the comparative analysis of the images in 2000 and 2013, it was found that the boundary density and patch shape index decreased, which indicates that the channel shape tended to become regular with the number of low-grade river channels. It is worth noting that the river-concentrated area is also the cultivated land concentrated area, which reflects the interdependence of water landscape and cultivated land. It's different from the mutually exclusive of construction land and rivers.

Table 14.6 Analysis of land use landscape in HPHR

Year	NP	MPS	ED	MPFD	MSI	SDI	SEI
1987	716	33.17	113.91	1.31	1.83	1.52	0.85
2000	2810	8.44	192.72	1.56	30.39	1.7	0.87
2013	2922	8.12	172.89	1.54	27.77	1.76	0.9

14.3.4 Changes of Landscape Pattern at Landscape Level

The change of landscape pattern's indices in HPHR is shown in Table 14.6. It is shown that the number of regional patches of NP increased sharply from 1989 to 2000, and decreased slightly after 2000. The comparison of the number of patches at type level shows that the main reason for this phenomenon is that the agricultural land and water area first appeared fragmentation during the period from 1989 to 2000, and then as the fragmented small patches were replaced by urban land, the number of agricultural land and water area landscape patches decreased.

The boundary density and patch shape index increased significantly from 1987 to 2013 and then decreased slightly, indicating that the patch shape tends to be more regular after the complexity caused by the initial landscape fragmentation. In terms of Shannon diversity and Shannon evenness indices of the regional landscape. They showed a gradual rise in trend which indicates that the land use types were gradually enriched, the fragmentation degree of patches increased and the distribution was relatively uniform from 1987 to 2013. This result is related to the different resolution and imaging time of remote sensing images in different periods.

The evolution process of land use and landscape pattern in the HPHR is jointly restricted by economic and natural ecological factors. The landscape pattern of land use in the initial stage is mainly affected by the industrial and agricultural production layout, with economic factors as the leading factor and natural factors as the second factor. From the analysis of land use landscape data. It can be seen that land use and ecological landscape change in the HPHR can be divided into two stages. Before 2000, affected by the accelerated process of economic and social development, the impact of construction land has become more and more important, the area of forest land and water area has shrunk significantly. This resulted in the degree of regional landscape fragmentation has become higher, and the ecological environment has a trend of deterioration; However, since 200, with the implementation of the Hanjiang Wetland Ecological Management Project, ecological land conversion to nonecological land showed a weakening trend after, while ecological quality improved. The richness of regional landscape patch types in the study area has increased and the landscape heterogeneity also increased. The ecological regulation function of the HPHR has been strengthened to a certain extent.

14.4 Conclusion and Discussion

Analysis of LUCC and landscape pattern can provide important data and decision support for ecosystem management. In this study, we used high-resolution long time series (1987, 2000 and 2013) remote sensing images to analyze LUCC in HPHR, and emphasized the landscape pattern change process of HPHR during the research period. The results of this study can provide reference information for water resource management and ecological protection in the HPHR.

In general, the land use types were similar for all the three periods, while the landscape diversity has declined. Cultivated land, construction land and woodland are the main land use types in the study area and conversion frequently occurred between each three land use types from 1980 to 2020. The land use and ecological landscape change in the HPHR can be divided into two stages before 2000 and after 2000: Before 2000, affected by the accelerated economic and social development process, the impact of construction land is becoming more and more important, the degree of regional landscape fragmentation is increasing, and the ecological environment has a trend of deterioration; After 2000, with the implementation of the Hanjiang Wetland Ecological Management Project, the richness of regional landscape patch types has increased, and the landscape heterogeneity has increased. The ecological regulation function of the HPHR has been strengthened to a certain extent.

Our findings show that the evolution process of land use and landscape pattern in the HPHR is jointly restricted by economic and natural ecological factors. The landscape pattern of land use in the early stage is mainly affected by the industrial and agricultural production layout, with economic factors as the leading factor and natural factors as the second factor. From an overall perspective, there are some problems in the landscape pattern of the HPHR: Human activities cause the decline of landscape connectivity; The construction of land patches is mainly concentrated in the flat areas on both sides of the main stream, and there are few on both sides of the tributaries. The ecosystem service value of the whole region has not been fully reflected. In view of the existing problems, the regional landscape pattern should be optimized, the ecological corridor should be constructed, the continuity between patches should be enhanced, and the weak ecological function on the ecological corridor should be identified as the ecological node to provide a reasonable landscape pattern model for the regional planning of the HPHR.

References

Cai SH, Hu X (2008) River ecosystem health and its evaluation. Adv Sci Technol Water Resour 28(1):23–27

Cao XY, Kou XM, Zhou MT (2022) Study and assessment on river health in the Pingchuan Section of the Hanjiang River in Hanzhong City. Northwest Hydropower 03:34–38+44

Chen WB, Xiao DN, Li XZ (2002) Classification, application, and creation of landscape indices. Chin J Appl Ecol 13(1):121

Chen L, Ren CY, Zhang B, Li L, Wang ZM, Song K (2018) Spatiotemporal dynamics of coastal wetlands and reclamation in the Yangtze Estuary during past 50 years (1960s–2015). Chin Geogra Sci 28(3):386–399
Chen X, Zhang ML, Zhang WC (2022) Landscape pattern changes and its drivers inferred from salt marsh plant variations in the coastal wetlands of the Liao River Estuary, China. Ecol Indic 145:109719
Corry RC (2005) Characterizing fine-scale patterns of alternative agricultural landscapes with landscape pattern indices. Landsc Ecol Eng 20:591–608
Feng SR, Guo L, Li DD, Huang Q (2018) Spatial patterns of landscape change in the Three Rivers Headwaters Region of China, 1987–2015. Acta Ecol Sin 38:76–80
Huang MS, Du M, Li P, Guo Y, Wang Y (2019) Investigation of the abrupt changes in precipitation concentration and the driving forces under the changing environment. Adv Water Sci 30(04):496–506
Kang LJ, Jang WW, Li J (2015) Study on optimization of ecological green space network in Changhe Town based on landscape pattern analysis. J Northwest For Univ 30(01):245–250
Khalyani AH, Mayer AL (2013) Spatial and temporal deforestation dynamics of Zagros forests (Iran) from 1972 to 2009. Landsc Urban Plan 117:1–12
Mcgarigal K, Marks BJ (1995) FRAGSTATS: spatial pattern analysis program for quantifying landscape structure (version 2.0). Forest Science Department, Oregon State University, Corvallis, 351
McGarigal K, Cushman SA, Ene E (2012) FRAGSTATS v4: spatial pattern analysis program for categorical and continuous maps. Computer software program produced by the authors at the University of Massachusetts, Amherst
Odongo VO, Oel PV, Christiaan V, Su Z (2019) Impact of land use and land cover transitions and climate on evapotranspiration in the Lake Naivasha Basin, Kenya. Sci Total Environ 682:19–30
Qin H, Chen Y (2023) Spatial non-stationarity of water conservation services and landscape patterns in Erhai Lake Basin, China. Ecol Indic 146:109894
Shirmohammadi B, Malekian A, Salajegheh A, Taheri B, Azarnivand H, Malek Z, Verburg PH (2020) Impacts of future climate and land use change on water yield in a semiarid basin in Iran. Land Degrad Dev 31:1252–1264
Song S, Liu Z, He C, Lu W (2020) Evaluating the effects of urban expansion on natural habitat quality by coupling localized shared socioeconomic pathways and the land use scenario dynamics-urban model. Ecol Ind 112:106071
Su BQ, Fang QW, Cheng ZY (2008) Discussion on flood control and risk removal technology for the main stream of Hanzhong Pingchuan in the upper reaches of Hanjiang River. Shaanxi Water Resour 03:67–68
Sunsanee A, Rajendra S (2016) Assessing land use change and its impact on ecosystem services in Northern Thailand. Sustainability 8(8):768
Wang GS, Xu WB, Lin QT (2006) Progress on evaluating methods for river health. Saf Environ Eng 13(4):14–17
Wang SQ, Zhou ZH, Liu DY, Hu Q, Tang WH, Zhou M (2022) Study on landscape pattern change of Chaohu basin based on land use change. Agric Sci J Yanbian Univ 44(04):84–93
Wu AN (2005) Study on river health assessment and its development in river management. East China Normal University
Wu AN, Yang K, Che Y (2005) Characterization of rivers health status and its assessment. Adv Water Sci 16(4):602–608
Xia HJ, Kong WJ, Zhou G, Sun JX (2021) Impacts of landscape patterns on water-related ecosystem services under natural restoration in Liaohe River Reserve, China. Sci Total Environ 792:148290
Yang WH (2007) Study on the theoretical framework and diagnostic system of river health. Hehai University

Chapter 15
Smart Monitoring of Constructed Wetlands to Improve Efficiency and Water Quality

Henrique Pinho, Manuel Barros, André Teixeira, Luís Oliveira, Pedro Matos, Carlos Ferreira, and Dina Mateus

Abstract The Smart monitoring of constructed wetlands to improve efficiency and water quality (SmarterCW) project aims to monitor biological wastewater treatment processes by gathering continuous data from remote water and environmental sensors. The acquired data can be processed and analyzed through data science tools to understand better the complex and coupled phenomena underneath wastewater treatment and monitor and optimize the system performance. The results will improve the efficiency and control of nature-based wastewater treatment technologies. The methodology comprises the following tasks and activities: Implementation of a set of electrochemical sensors in the input and output flow streams of pilot-scale constructed wetlands; Acquisition of water quality parameters such as pH, electrical conductivity, temperature, and ionic compounds; Acquisition of environmental parameters, such as temperature and humidity; Application of data analysis tools to design and optimize conceptual models to correlate pollutants removal with operative parameters in green technologies for wastewater treatment. This methodology was applied to a patent-protected pilot-scale modular constructed wetland in which filling media consists of a mixture of solid waste. A high-level IoT communication layer structure complements the system to support remote real-time water and environmental monitoring, system performance, and data dissemination.

Keywords Wastewater treatment · Sensor networks · Efficiency monitoring

15.1 Introduction

Constructed wetlands (CW) consist of a nature-based solution (NBS) for wastewater treatment (Cross et al. 2021). NBS technologies are greener solutions aiming to drive

H. Pinho (✉) · M. Barros · A. Teixeira · L. Oliveira · P. Matos · C. Ferreira
Ci2—Smart Cities Research Center, Instituto Politécnico de Tomar, Tomar, Portugal
e-mail: hpinho@ipt.pt

D. Mateus
Techn&Art—Technology, Restoration and Arts Enhancement Center, Instituto Politécnico de Tomar, Tomar, Portugal

H. Xu (ed.), *Proceedings of the 5th International Symposium on Water Resource and Environmental Management*, Environmental Science and Engineering,
https://doi.org/10.1007/978-3-031-31289-2_15

sustainable development (Bunclark and Hernández 2022). CW are low-cost, ecological, and effective wastewater treatment systems (Vymazal 2022). CW can be applied to treat diverse types of wastewaters, including domestic (Rosendo et al. 2022), industrial (Vymazal 2014), and greywater (Arden and Ma 2018) among others, targeting different kinds of pollutants including conventional pollutants such as nutrients, but also specific compounds such as steroids and antibiotics (Kamilya et al. 2023), and heavy metals and metalloids (Yu et al. 2022), among others. Furthermore, CW can contribute to circular economy of water through the production of reclaimed water (Pinho and Mateus 2022), valorization of solid waste as filling media (Mateus and Pinho 2020), and valorization of the wetland plants for bioenergy applications (Pinho and Mateus 2023).

There are several types of CW, but the common types comprise three main components (Parde et al. 2021):

1. a water retention structure, such as a pond lined with an impervious membrane or a kind of container
2. a bed of granular filling material with adequate hydraulic permeability, such as gravel or solid waste
3. water-tolerant plants such as the macrophyte common reed.

However, these technologies have some disadvantages related to a minor level of control than conventional technologies. As an example of a conventional solution, an anaerobic reactor treating wastewater can be operated at different flow rates, temperatures, and mixing intensities. Manipulating those parameters can adapt the process efficiency to changes in the wastewater composition. On the other hand, NBS such as CW has limited level of manipulation. Although some variants of CW can have notable improvements, such as tidal flow operation or bed aeration, in the most common CW systems, only the flow rate can be varied but to a low extend over the average designed value (Vymazal et al. 2021). Thus, water quality parameters should be monitored at the higher available rate to optimize CW operation by the dynamic setting of the input wastewater flow rate. CW and NBS technologies commonly involve discrete water sampling and further analysis in a laboratory at a low frequency, such as daily, weekly, or even monthly sampling (Mateus and Pinho 2020).

In the SmarterCW project, a set of sensors for continuous evaluation of the water quality aims to get insights on the system response to the change of the uncontrolled environmental variables, such as air temperature, humidity, speed, radiation intensity, and pluviometry, among others, and design a framework for future implementation of back control and optimization models.

The goal of this work consists of reporting the preliminary stages of the sensor network framework development based on physical water parameters, and presenting examples of the gathered data during a representative time of operation.

Fig. 15.1 Picture of the CW prototype (**a**), and schema showing the filling media consisting in four layers of three recovered solid wastes (**b**)

15.2 Material and Methods

15.2.1 CW Prototype

The CW prototype is fully described in previous works (Pinho and Mateus 2021). The prototype consists of a modular planted bed filled with a mixture of solid waste, treating secondary-type domestic wastewater (Fig. 15.1). The bed has a surface area of 1.1 m^2 and a depth of 0.5 m, planted with common reed (*Phragmites australis*). The water flows horizontally 0.1 m below the surface, with an average load of 40 L/day. The filling consists of four layers of three solid waste: a bottom layer and a top layer of fragmented limestone rock, a residue from construction activities; a layer of coal slag from a coal power plant above the bottom layer, and a layer of cork granulates from the cork industry, over the coal slag layer. Besides acting as a support for the plants and microorganisms community, the different solids contribute to wastewater treatment. The cork residues and coal slag play a role as sorbents for organic pollutants, and the limestone residues adsorb phosphorous compounds, a recalcitrant nutrient in wastewaters (Mateus and Pinho 2018).

15.2.2 Sensor Network

Preliminary works are carried out using two sets of three sensors from Aqualabo (France): pH probe PHEHT, Dissolved Oxygen (DO) probe OPTOD, and Electrical Conductivity (EC) probe C4E, and two sets of integrated pH, EC, and Temperature probes (Groline HI 981420, Hanna Instruments, Germany). One set is located in the input flow, and one in the output flow of the CW prototype. The integrated Groline sets function as reference for the data acquisition system.

Figure 15.2 presents the main framework for data acquisition and processing from the probes installed in the input and output streams of the CW prototype. The feedback from the input and the control system's output stream is a fundamental component to satisfy the requirements of a fully automated data acquisition system. The implemented hardware and communication software was selected based on optimizing costs but providing scalability for easy addition of more probes in the future.

Two ESP32 microcontrollers were used for the acquisition of water quality parameters from the sensors and controllers using the interface RS485 and are in charge for the implementation of two additional functions: on the input side, the ESP is responsible for the management of a peristaltic pump, the device used to control the variable of the close-loop systems. The ESP32 on the output side also has the additional function of controlling the discharge of water from the tank at the system outlet. The control information can be generated locally, through a Raspberry Pi (RPi) microprocessor or by the central monitoring platform software developed for the cloud. The RPi microcontroller also performs the functions of a gateway device providing connectivity, security, and data routing, between the private application network and the public Internet. The publish-subscribe standard message protocol

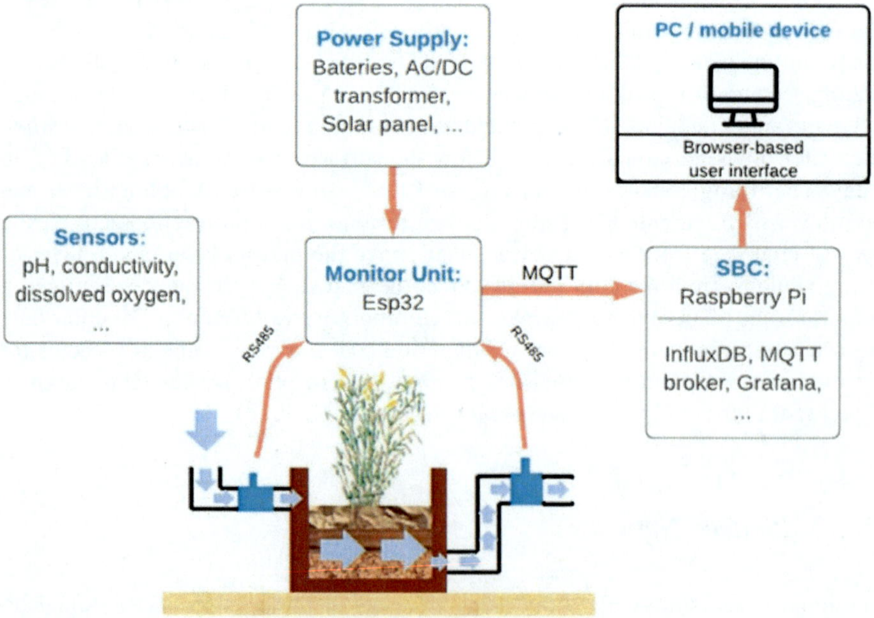

Fig. 15.2 Scheme of the implemented sensor network in the CW prototype. SBC represents a Single Board Computer such as the Raspberry Pi used in this work

MQTT (Message Queue Telemetry Transport) was used for local communications because is well suited for bidirectional data transmission from many local devices to a single central monitoring station. Finally, a cloud-based platform was developed to present data to the user and provide ways to interact with the application. This application provides a user-friendly interface for accessing, storing, controlling, and presenting the application parameters. The software package is designed to implement analytic algorithms to analyze time-series data and provide system insights, enabling the monitoring system to be more effective and efficient over time. The system was also designed to interact with Grafana and InfluxDB open-source monitoring tools.

15.2.3 Efficiency Evaluation

In this work, the efficiency of wastewater treatment is computed using the Electrical Conductivity results. Although pH and DO help understand the physicochemical and biological phenomena that occur in CW, the EC is an indicator of the concentration of dissolved substances. It thus can be a reasonable indicator of the efficiency of wastewater treatment.

The treatment efficiency is computed from the readings of the EC probes in the input and output streams, according to Eq. (15.1).

$$\eta = (EC_{in} - EC_{out})/EC_{in} \times 100 \tag{15.1}$$

where EC_{in} and EC_{out} are the readings of the probes in the input and output streams, respectively, and η the treatment efficiency (%).

15.3 Results and Discussion

Figure 15.3 shows an example of the gathered data for a typical run of 4 days, with a probe sampling rate of 15 min. The output is strongly irregular, as already reported for CW facilities (Song et al. 2022), showing a cyclical daily variation. The minimum EC values occur daily between 2 and 4 pm, and the maximum values occur daily between 10 and 12 pm. Minimum EC values imply higher dissolved solids removal efficiency. These results may be related to higher temperatures and solar radiation, improving the pollutants assimilation by the plants and by the microbiological community colonising the CW bed.

Considering a traditional sampling procedure, for example, by 10 am every day, the computed treatment efficiency is $9 \pm 3\%$. However, computing the treatment efficiency using all the gathered data, the average efficiency is $7.2 \pm 0.3\%$. Both differences in the average value and uncertainty interval are significant. In this example, a low-frequency sampling resulted in a 20% overvaluation of the treatment efficiency.

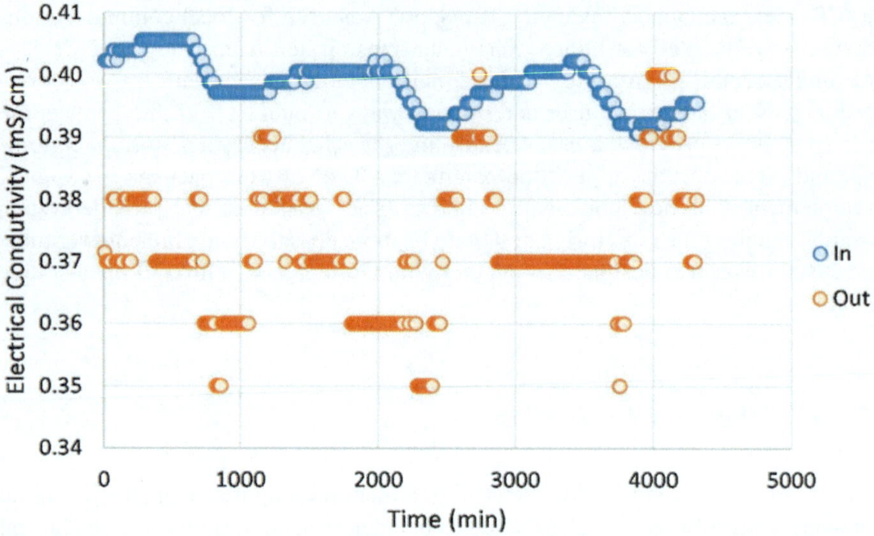

Fig. 15.3 Example of registered EC in the input and output water streams in the CW prototype, obtained by the integrated set of sensors

As previously introduced, CW, like other NBS systems, have a low degree of manipulation during operation. The most controllable parameter is the feed flow rate of wastewater. The continuous knowledge of the treatment efficiency can optimize the operation by maneuvering the input flow rate, increasing when the treatment efficiency increases, and reducing the flow rate when the treatment efficiency decreases. Moreover, the real-time monitoring of other variables, including environmental parameters, can further increase the capacity to adjust the input flow rate, making these systems more reliable and sustainable.

Presently, besides EC, also pH and DO probes are installed and their data is continuously gathered. The daily variation of pH and DO is presented in Figs. 15.4 and 15.5, respectively. As for EC, the daily variation of pH and DO is cyclical, with special evidence on the DO behavior. The evaluation of these parameters only through low-frequent samplings can, also, present significant deviations to the average values, and lead to a deficient interpretation of the behavior of the CW.

Current work is ongoing to install more sensors, such as COD, nitrate, ammonium, and phosphate probes. Those parameters present better understanding of CW efficiency than pH, EC, and DO, thus the future availability of them will allow the developing of models that can contribute to optimize CW operation. In addition, the sensor network under development can easily be applied to full-scale CW in real-field operation.

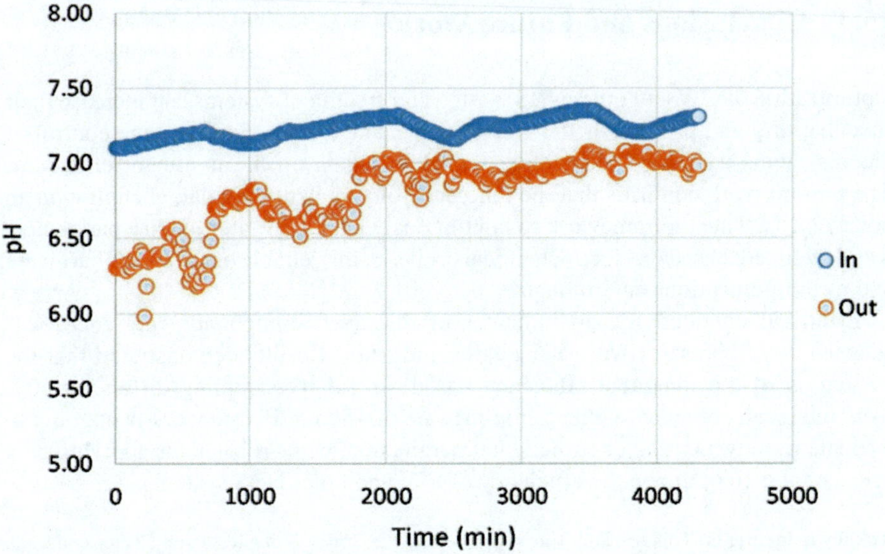

Fig. 15.4 Example of registered pH in the input and output water streams in the CW prototype, obtained by the integrated set of sensors

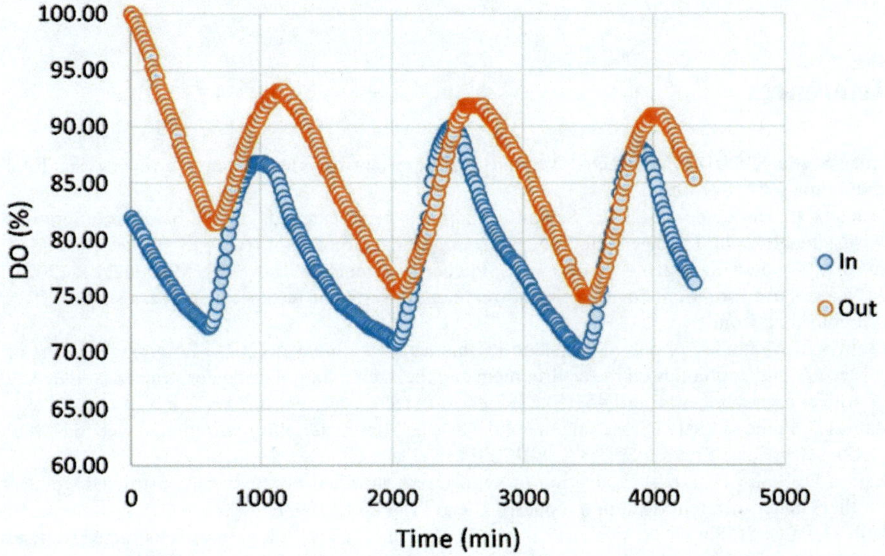

Fig. 15.5 Example of registered DO in the input and output water streams in the CW prototype, obtained by the integrated set of sensors

15.4 Conclusions and Future Work

Optimization of CW and other NBS wastewater treatment systems can increase their sustainability and promote better freshwater management. However, the control of those systems is limited, depending on reliable data on the treatment efficiency. The present work confirms that the efficiency of CW is not regular, changing on an hourly basis. Thus, a framework to continuously evaluate water quality parameters was developed, based on electrochemical probes using reliable but low-cost hardware and a communications environment.

Electrical conductivity, an indicator of dissolved compounds' presence, was selected as a representative water quality indicator. Results demonstrated that the evaluation of the treatment efficiency based on a daily sampling differs by 20% from the results obtained with a sampling rate of 15 min. Future work is ongoing to evaluate more water and environmental parameters, being a basis for applying data science tools to optimize the efficiency of CW and other NBS systems.

Acknowledgements This research was supported by the Smart Cities Research Center under the grants UIDB/05567/2020 and UIDP/05567/2020 funded by The Portuguese Foundation for Science and Technology (FCT—Fundação para a Ciência e a Tecnologia, Portugal). The grant POCI-01-0145-FEDER-23314 is also acknowledged.

References

Arden S, Ma X (2018) Constructed wetlands for greywater recycle and reuse: a review. Sci Total Environ 630:587–599

Bunclark L, Hernández I (2022) Scientific mapping of research on nature-based solutions for sustainable water management. Water Resour Manage 36:4499–4516

Cross K, Tondera K, Rizzo A, Andrews L, Pucher B, Istenič D, Karres N, McDonald R (2021) Nature-based solutions for wastewater treatment: a series of factsheets and case studies. IWA Publishing, London

Kamilya T, Yadav M, Ayoob S, Tripathy S, Bhatnagar A, Gupta A (2023) Emerging impacts of steroids and antibiotics on the environment and their remediation using constructed wetlands: a critical review. Chem Eng J 451(3):138759

Mateus D, Pinho H (2018) Screening of solid waste as filler material for constructed wetlands. IOP Conf Ser: Earth Environ Sci 182(1):012001

Mateus D, Pinho H (2020) Evaluation of solid waste stratified mixtures as constructed wetland fillers under different operation modes. J Clean Prod 253:119986

Parde D, Patwa A, Shukla A, Vijay R, Killedar D, Kumar R (2021) A review of constructed wetland on type, treatment and technology of wastewater. Environ Technol Innov 21:101261

Pinho H, Mateus D (2021) Valorization of solid waste in subsurface flow constructed wetlands based on renewable modular structures: a contribution to a circular economy. In: Circular economy and sustainability. Environmental engineering, vol 2. Elsevier, Amsterdam, pp 215–233

Pinho H, Mateus D (2022) Contribution of constructed wetlands for reclaimed water production: a review. IOP Conf Ser: Earth Environ Sci 1006(1):012008

Pinho H, Mateus D (2023) Bioenergy routes for valorizing constructed wetland vegetation: an overview. Ecol Eng 187:106867

Rosendo J, da Paz G, Rosendo A (2022) Constructed wetlands applied on domestic wastewater for decetralized systems: concepts, processes, modalities, combinations and enhancements; a review. Rev Environ Sci Biotechnol 21:371–397

Song S, Sheng S, Xu J, Zhao D (2022) What is the suitable sampling frequency for water quality monitoring in full-scale constructed wetlands treating tail water? Water 14:2431

Vymazal J (2014) Constructed wetlands for treatment of industrial wastewaters: a review. Ecol Eng 73:724–751

Vymazal J (2022) The historical development of constructed wetlands for wastewater treatment. Land 11(2):174

Vymazal J, Zhao Y, Mander Ü (2021) Recent research challenges in constructed wetlands for wastewater treatment: a review. Ecol Eng 169:106318

Yu G, Wang G, Chi T, Du C, Wang J, Li P, Zhang Y, Wang S, Yang K, Long Y, Chen H (2022) Enhanced removal of heavy metals and metalloids by constructed wetlands: a review of approaches and mechanisms. Sci Total Environ 821:153516

Chapter 16
Experimental Study on Dehydration of Dredged Sediment from Urban Inland Rivers and Lakes and Analysis of Economic Benefits

He Bai, Jiachen Zeng, Bing Guo, Yuchi Hao, Xiaowei Yan, and Zheng Lu

Abstract In this paper, in view of the single dosing type, environmental protection and high cost problems existing in the traditional river and lake dredging sediment treatment project. We takes the typical urban river and lake dredging project as the research object, analyzes the dewatering effect of multiple agents added separately or in combination on the river and lake sediment through the sludge specific resistance, SV_{30} sedimentation and small and medium simulation press tests, expounds the deep dewatering mechanism of different agent types selected, The economic benefits of different chemicals on the project construction were analyzed according to the dosage and cost of chemicals. Results show that under the condition that the dehydration rate is lower than 40% and the tail water reaches the standard, the organic polymer agent NEM has better economic benefits, dehydration rate is lower than 35% and the compressive strength is around 50 Mpa, which is better than the compound agent OPF + GSS with the same treatment effect in cost.

Keywords Simulation extrusion · Sludge · Dehydration effect · Sludge specific resistance

16.1 Introduction

River and lake bottom mud is an important part of the water ecosystem and an important reservoir of various pollutants in rivers and lakes (Yang et al. 2020). With the rapid urban development, a large number of pollutants are collected in the bottom mud through the processes of adsorption, complexation, flocculation and sedimentation, resulting in increasingly serious water pollution (Zhong et al. 2008). Ecological dredging of bottom mud (Piló et al. 2019), as a technical method that can effectively remove endogenous pollutants from rivers and lakes and improve water quality, produces dredged bottom mud with high mobility, high water content, diverse mud

H. Bai (✉) · J. Zeng · B. Guo · Y. Hao · X. Yan · Z. Lu
CCCC National Engineering Research Center of Dredging Technology and Equipment Co., Ltd, Shanghai 200082, China
e-mail: baihe@cccc-drc.com

H. Xu (ed.), *Proceedings of the 5th International Symposium on Water Resource and Environmental Management*, Environmental Science and Engineering,
https://doi.org/10.1007/978-3-031-31289-2_16

quality and complex pollutants that make it impossible to be directly transported out and used resourcefully (Ding et al. 2015; Liu et al. 2020). Therefore, it is necessary to vigorously carry out the research on bottom mud drying to effectively prevent the secondary pollution of dredged bottom mud and further realize the resource utilization of river and lake bottom mud (Kasmi et al. 2017).

So far, flocculation dewatering technology (Xu et al. 2022; Cao et al. 2021) among dredged bottom drying technologies has increasingly become the focus of key pretreatment measures for dredged bottom due to its simple operation method, fast settling rate and low drying cost (Wei et al. 2018). Pharmaceutical agents are the core of flocculation and drying technology, which mainly promote the destabilization of the system through the electro-neutralization of the charged ions carried with the sludge surface charge, the compressed double layer effect, and the co-precipitation effect of the polymeric flocculant net catching and sweeping to accelerate the sludge particles aggregation and settlement, thus achieving the effect of mud-water separation (Li et al. 2018), and their type, dosage and other parameters conditions directly affect the treatment effect of bottom sludge dewatering (Jiang et al. 2021). However, the current research on pharmaceuticals often focuses on the dewatering mechanism (Lu et al. 2019), but there is little analysis on how to select pharmaceuticals (Mikkelsen et al. 2018) and whether they can be used in engineering applications, and for projects with special requirements, such as substrate dumps and resource recovery, the effects of pharmaceuticals on substrate bearing strength, pH, and nutrient composition need to be studied (Sun et al. 2020; Guo et al. 2019; Onoguchi et al. 2019). In addition, since the types of pharmaceutical agents include organic, inorganic, organic–inorganic compound, biological agents, etc. (Pandey et al. 2019), different pharmaceutical agents may have certain pollution risks to the environment, and the negative effects of different pharmaceutical agents on the environment have not been sufficiently studied, so the selection of pharmaceutical agents needs to be fully tested and environmentally evaluated to confirm that they meet the relevant national environmental standards before use (Bo 2020).

Based on these problems, this study investigates the dewatering effect of different types of chemicals alone and in combination on dredged bottom sludge in urban lakes and rivers, analyzes the effect through sludge specific resistance test, settlement test and lateral limitless compressive strength test, and analyzes the relationship between sludge quality and types of chemicals, and verifies its feasibility in the process of engineering application through small test and pilot plate and frame filter press test, so as to explore the suitable chemical solution for dewatering dredged sludge in urban lakes and rivers, and conducts benefit analysis of the solution to provide reference opinions for the targeted selection of chemicals in actual projects.

16.2 Materials and Methods

16.2.1 Materials

Dredging sediment. The test mud was taken from a typical river and lake in Wuhan, and the sampling location was at the outlet of the dredging pipeline. After multiple sampling, the mud physical property indexes were tested, and the test results are shown in Table 16.1.

Reagent type. The agents selected for the test were from XX agent manufacturers, including organic agents, inorganic agents and other types of agents, with differences in molecular weight between the agent types. The agents were diluted in different proportions, the density of the diluted agents was tested, and the dewatering test study of the dredged sediment was conducted with the diluted agents. The details are shown in Table 16.2.

16.2.2 Dehydration Test

The dewatering test of dredged bottom mud mainly includes sludge specific resistance test device, sludge settling test (SV30), sludge simulation press test, small plate and frame filter press test and lateral limitless compressive strength test.

Table 16.1 Physical properties of dredged sludge

Soil sample	Density (g/L)	Water content (%)	pH	Organic matter content (mg/L)
Dredged sludge	1.20	19.98	8	800

Table 16.2 Test parameters of various reagents

Name	Molecular weight	Dilution ratio	Density after dilution (g/L)
OPF	1500 w	1000	1.05
NEM	200 w	10	1.08
NMS	200 w	10	1.02
NRL	200 w	10	1.10
NLS	200 w	10	1.03
OP	1200 w	1000	1.02
BA	1200 w	1000	1.03
IA	1200 w	10	1.04
GGS	–	0	–

Table 16.3 Parameters of sludge specific resistance test

Mud mixing time (min)	Mud volume (g)	Suction pressure (MPa)	Test paper model
5	100	0.8	9

Sludge specific resistance test. During the test, take 100 g of bottom sludge with 19.98% solids in a beaker, add a certain proportion of diluted chemicals, stir at a constant speed for 5 min, pour the stirred mud-water mixture into a Brinell funnel, turn on the vacuum pump for 15 min or the pressure gauge starts to drop, then take out the filtered product, test the water content of the product and calculate its sludge specific resistance.

The sludge specific resistance test parameters are shown in Table 16.3.

Sludge settling test. Take 100 ml of bottom mud containing 19.98% solids in a beaker, add a certain proportion of diluted chemicals, stir at a constant speed for 5 min, pour the mixed and processed mud mixture into a 250 ml measuring cylinder quickly, let it stand for 30 min and then take readings to calculate the settling rate.

Simulated press to verify sludge cake effect. The mud-water mixture was poured into the press barrel and pressed for 20 min under 0.8 MPa pressure by adjusting the air pressure valve. The parameters of the simulated pressing test are shown in Table 16.4.

Small filter press pilot test. Take 12.5 kg of bottom mud with 19.98% solids in the feeding device, add a certain proportion of chemicals, stir at a constant speed for 10 min, pump to the press device under low pressure of 0.6 MPa, press under 0.8 MPa pressure for 20 min, then take out the press product and test the water content of the product. The test parameters of the small-scale filter press are shown in Table 16.5.

Table 16.4 Parameters of simulated extrusion test

Slurry mixing time (min)	Slurry volume (g)	Pressing pressure (MPa)
5	200	0.8

Table 16.5 Test parameters of small filter press

Mixing time (min)	Filter plate pressure (MPa)	Feed pressure (MPa)	Feeding time (min)	Pressing time (min)	Pressing pressure (MPa)
10	0.8	0.6	40 min	20 min	0.8

16.3 Results

16.3.1 Sludge Specific Resistance

Sludge specific resistance is a conventional method of sludge property testing, which refers to the impedance per unit mass of sludge per unit filtration area when filtering at constant pressure. Sludge specific resistance is an important and comprehensive index reflecting the good or bad sludge dewatering performance. The dewatering test was carried out for eight different chemicals to test the variation of their sludge specific resistance with the dosing concentration, and the sludge specific resistance curve was obtained as shown in Fig. 16.1.

It was found that the measured sludge specific resistance values of the above eight types of agents showed a decreasing and then increasing trend as the concentration increased, and the best agents were BA, OP, OPF, IA, NEM, NRL, NLS, and NMS, where NMS showed a decreasing trend, indicating that the filtration performance of the sludge was better with the increasing of this agent. The differences in the values taken can be attributed to the differences in molecular weights and types of agents in the various types of agents. For the sludge specific resistance measurements, it was found that most of the measured specific resistance values were below $9 * 10^9 (s^2/g)$, and according to the relevant literature, for the primary sludge, the specific resistance values ranged from 4.7 to $6.2 * 10^9 (s^2/g)$, while the specific resistance measurements for BA and OP agents were larger in the selected range, so it was speculated that for the substrate in this area, BA and OP are not suitable or the dosage of the agents needs to be increased.

16.3.2 SV_{30} Sinker

SV_{30} is the most commonly used test method for sludge settling performance, meaning the volume ratio of the mixture to which the sludge can settle, which can better reflect the degree of sludge loosening and coalescence settling performance. Based on the test results of sludge specific resistance determination, the SV_{30} settling test was conducted for five different types of agents to test their sludge settling within 30 min, and the SV_{30} curve was obtained as shown in Fig. 16.2.

From the above SV_{30} settling curve, it can be found that the settling performance of the five randomly selected agents is not good. The agent with the maximum settling performance in 30 min time is BA, which settles from 100 ml to about 60 ml with a settling amplitude of about 40 ml. the agent type with the worst settling performance is NMS, with a maximum settling amplitude of only about 20 ml. By corresponding to the test results of its sludge ratio resistance, it was found that the test results of sludge settling did not provide a reference value for the effect of the agent in the plate and frame machine, because the special characteristics of the substrate led to the agent acting more as a kind of conditioning, and it was not very obvious to

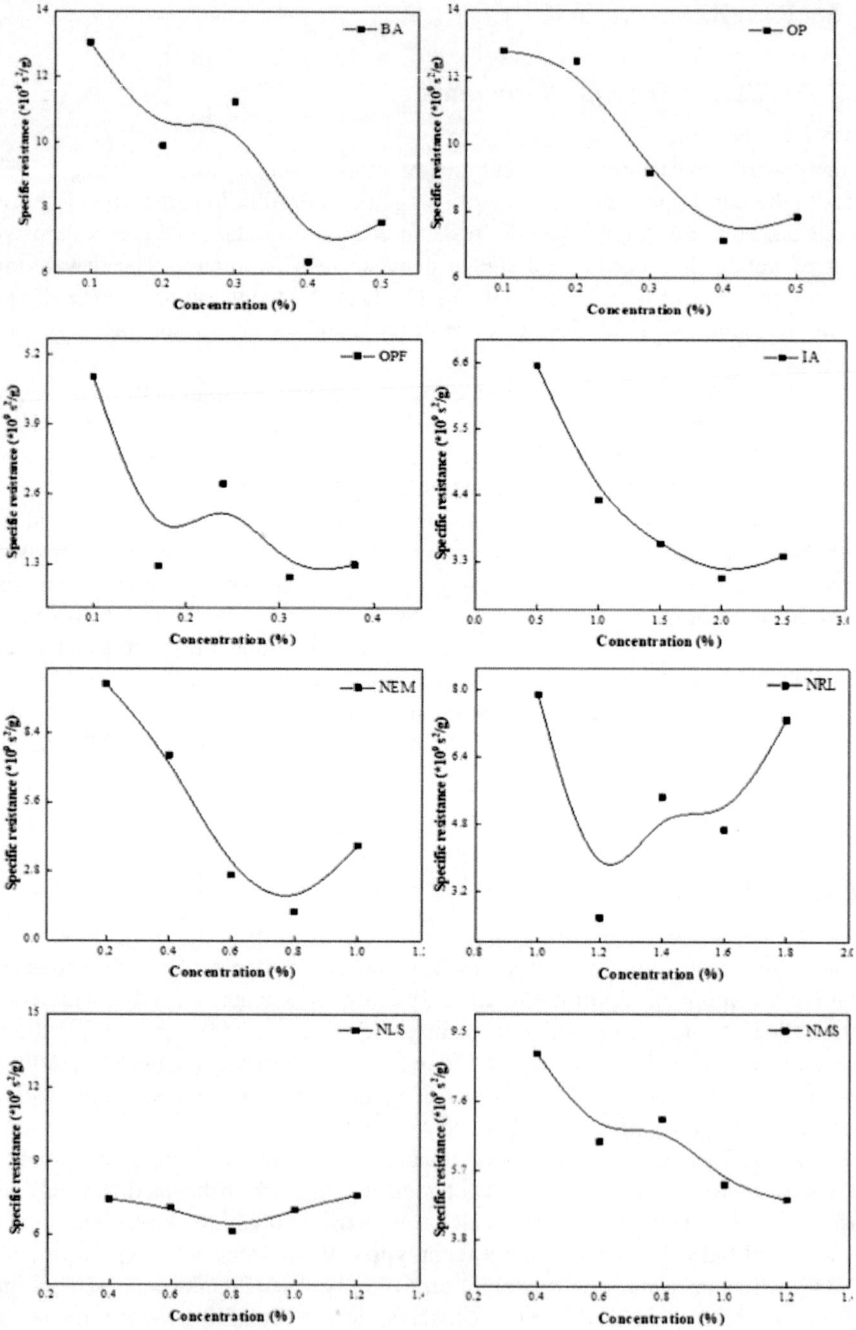

Fig. 16.1 Sludge specific resistance curve of different agents

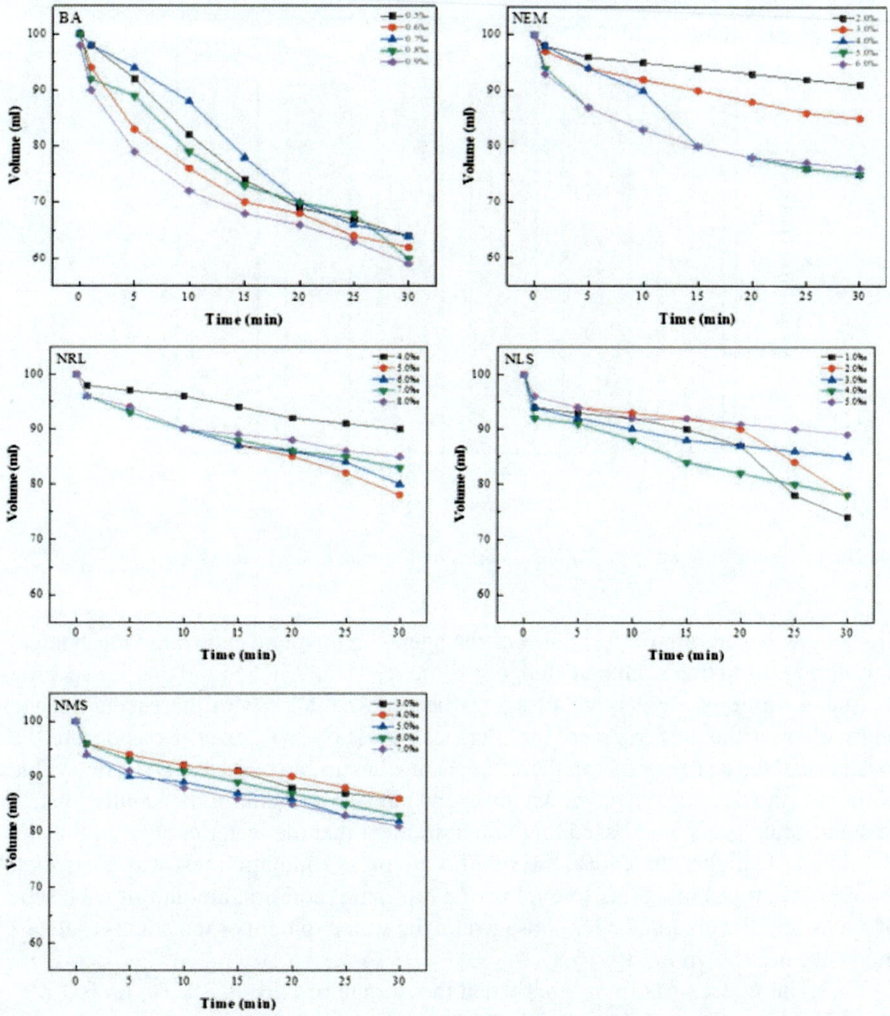

Fig. 16.2 SV_{30} settlement curve

flocculate the substrate and produce the phenomenon of mud-water separation, which was consistent with the phenomenon of the field test process.

16.3.3 Simulated Pressing and Cost Analysis

Small pilot mock press and cost analysis. The pilot test is an indoor test necessary to conduct the actual engineering study, in which the dewatering press simulation of the screened agents is performed under simulated actual working conditions to

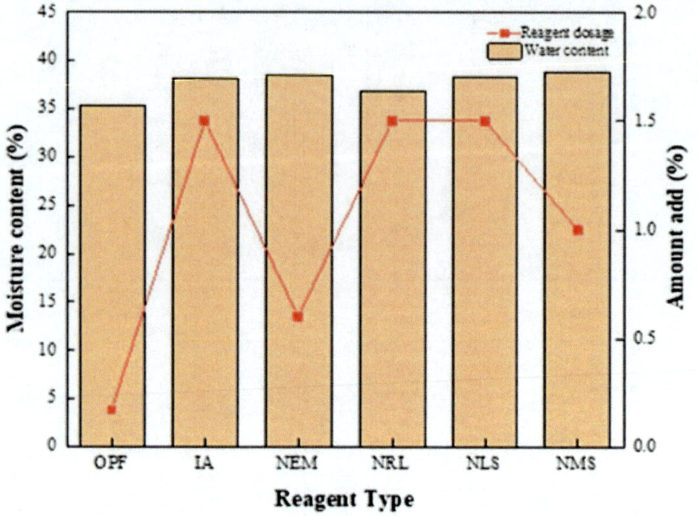

Fig. 16.3 Results of small scale simulation pressing test with different reagents

predict the dewatering performance of the agents in the plate and frame mechanical dewatering. From the results of sludge specific resistance, it is found that the specific resistance value of eight types of agents becomes smaller with increasing dosage and the better the performance, but a critical threshold will exist. Considering the existence of the test error factor, therefore, using the turning point as the starting value of the test and floating up and down the agent addition amount for the simulation test of small-scale press, and based on the requirement that the water content of the site should not be higher than 40%, the small-scale press simulation test was conducted for these six types of agents to explore the minimum addition amount of each type of agent, and the obtained agent dosage and the water content of the small-scale test mud cake are shown in Fig. 16.3.

From the figure, it is found that when the dosage of OPF is 0.17%, IA is 1.5%, NEM is 0.6%, NRL is 1.5%, NLS is 1.5% and NMS is 1.0%, the mud cake formed under the small test simulation press can reach the requirement of water content below 40%, and the water content of the mud cake is basically between 35 and 40%, where the water content of OPF The water content of OPF was 35.33%, IA was 38.2%, NEM was 38.5%, NRL was 36.9%, NLS was 38.4%, and NMS was 38.9%. This addition amount can be considered as the minimum amount. Under the condition of minimum dosage, the use of each type of agent in descending order is: OPF < NEM < NMS < IA = NRL = NLS.

The cost of each type of agent required to treat one cubic meter of submerged substrate with the same amount of agent addition as above is shown in Fig. 16.4.

The graph shows that GGS, OPF and IA have obvious price advantages, among which GGS is the curing agent used in a project, NRL, NLS and NMS are more expensive and not particularly suitable for industrial production. NEM is more expensive

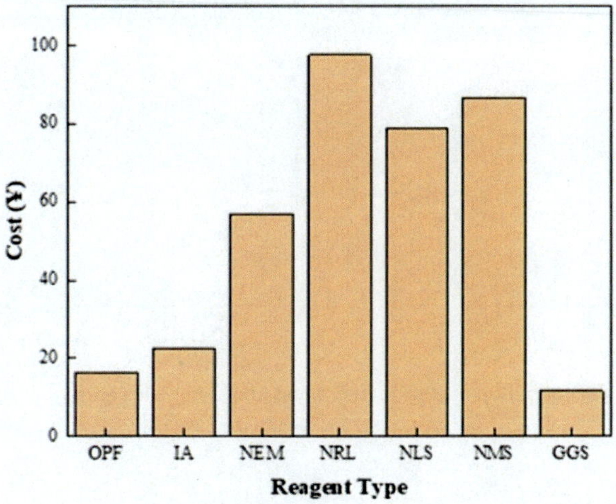

Fig. 16.4 Cost corresponding to the results of small scale simulation press test

but still within the control range, and its tailwater pH test is neutral, which can eliminate the post-treatment link of the tailwater. Based on the results of this cost analysis, four types of agents, GGS, OPF, IA and NEM, were tested in a pilot plate and frame filter press to test their dewatering effects after scaling up the process and to provide reference for practical industrial applications.

Pilot plate and frame filter press and cost analysis. Figure 16.5 shows the test results of the dewatering effect and the inlet and outlet mud volume of the four types of pure chemicals, GGS, OPF, IA and NEM, It is found that the water content of the mud cake after pressing is lower compared with that of the small test, which indicates that the actual dosage of chemicals can be reduced after the process is scaled up. From the curves of mud feed and discharge, it was found that NEM had higher mud feed and little change in discharge compared with others, which indicated that NEM had better bottom mud conditioning effect.

Figure 16.6 shows the dewatering effects and the inlet and outlet mud volumes of OPF, IA and GGS after compounding. From the figure, it can be found that the water content of the filtered mud cake gradually decreases with the increase of GGS addition, and the incoming mud volume is the most when the GGS addition is 1.5%, and the outgoing mud volume under different GGS addition conditions is basically the same. Therefore, it is considered that the optimal addition amount of GGS is 1.5%.

When the addition amount of GGS was fixed, the results of OPF, IA and NEM were tested for the water content of mud cake after filtration when the three types of agents were compounded with GGS, and the results are shown in Fig. 16.7. It is found in the figure that under the condition of water content close to 40%, the results of compounding of three types of agents can meet the requirements, and the effect of 0.13% OPF is far from the limit of 40%, which indicates that there is room for

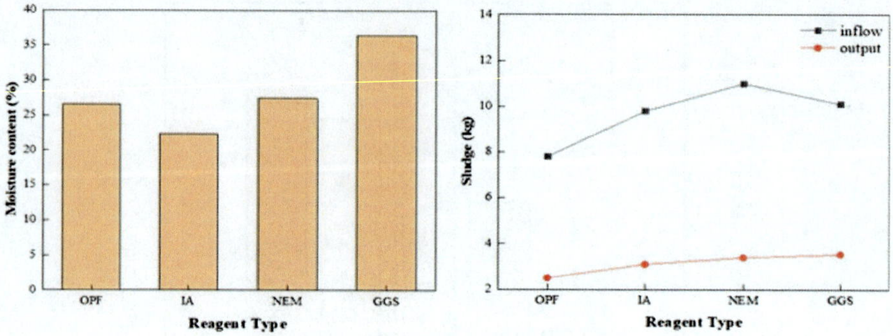

Fig. 16.5 Single reagent pilot plate frame filter press and sludge discharge

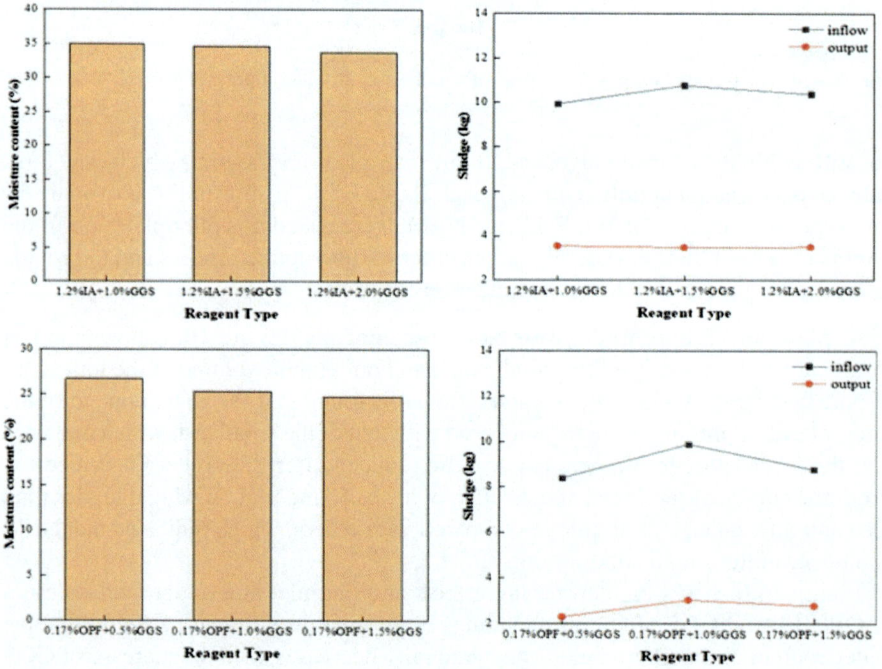

Fig. 16.6 Compression test of pilot plate frame mixed with GGS of different contents and sludge inflow and outflow

further reduction of OPF use to decrease the amount of mud in and out of the curve graph can be seen that the mud after the compounding of 0.13% OPF added can enter more mud under the working conditions of plate and frame filter press, out The mud volume is basically the same as the others, which indicates that the formulation has better dewatering effect.

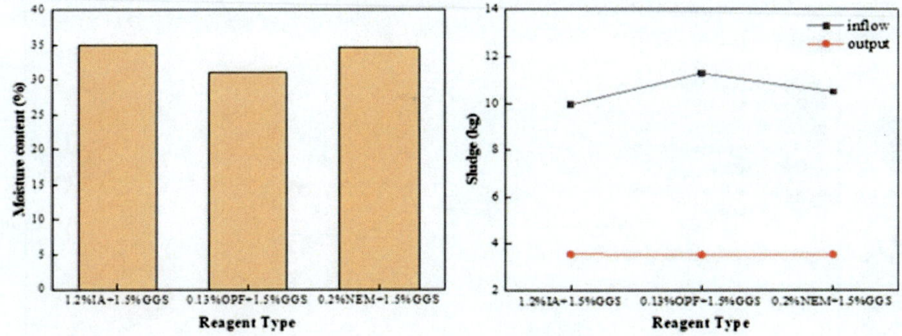

Fig. 16.7 Pilot plate and frame press test of GGS compounding with different agents and the inlet and outlet mud volume

16.3.4 Compressive Strength Without Lateral Limit

The difference of agent types is one of the main reasons affecting the strength of the dewatered mud cake after substrate dewatering. The test selected the dewatered mud cake obtained from different types of agent formulations for the comparison test of unconfined compressive strength, and the comparison graph of its strength is shown in Fig. 16.8.

From the figure, it can be found that the highest and most stable strength of the dewatered mud cake among all agent types is the 5% GGS agent, whose strength reaches between 50 and 55 kPa, which is related to the large amount of Ca+ in its composition. The lowest strength formulation was 0.17% OPF formulation, whose strength of dewatered mud cake was only between 22 and 24 kPa. Besides, the maximum strength of 0.2% NEM +1.5% GGS and 0.6% NEM formulations can also reach the effect of 5% GGS formulation, but the 0.2% NEM +1.5% GGS formulation is more volatile and the 0.6% NEM formulation is less volatile, which is mainly due to the chemical reaction between the Ca+ ions in GGS and some components in NEM, thus reducing the strength of the dewatered mud cake. The strength of the mud cake after dewatering. It can be seen that 0.6% NEM is better than 0.2% NEM +1.5% GGS in overall performance. 0.13% OPF +1.5% GGS formulation is slightly deficient in strength, and its fluctuation range is from 40 to 50 kPa.

16.4 Mechanistic Analysis

The agents used in this experiment are mainly divided into two categories: organic flocculants and inorganic conditioning agents, of which organic flocculants are mainly represented by OPF and NEM, and inorganic agents are mainly GSS and IA.

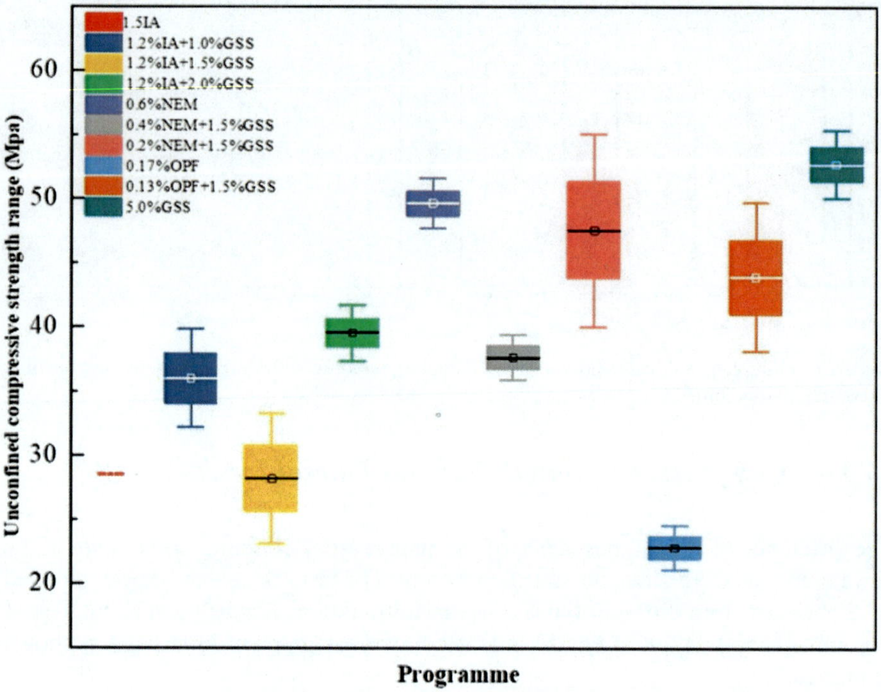

Fig. 16.8 Unconfined compressive strength of mud cake with different compounding schemes

There are two main mechanisms by which organic agents act on river and lake bottom mud dewatering.

Electro-neutralizing effect. That is, there is a double potential on the surface of the solid ions, the electrostatic material of the double layer due to the charge generated by the positive and negative ions repel each other and prevent them from colliding with each other, and the stable suspension in the liquid cannot settle, so adding a flocculant with the opposite charge of the solid example in the slurry, so that the double layer potential of the slurry particles drop, the solid particles contact each other and flocculate into large flocs and settle, the electric neutralization effect in this test The representative flocculant in this test is OPF.

Adsorption bridging effect. Polymer flocculant molecular weight, long chain, along the length of the chain there are many Mars functional groups, and therefore can adsorb multiple suspended solid particles. When the chain of polymer molecules in solution is adsorbed to a particle, if another particle is close, the polymer chain can be adsorbed to the particle, forming a bridge between the two particles. The basic flocculant continues to grow by bridging with other particles until a large floc settles.

Secondly, the mechanism of inorganic conditioning agents is mainly the precipitation of insoluble substances, and the inorganic conditioning agents mainly used in this test are GSS and IA. The principle is mainly due to the electrostatic interaction between the cationic groups of polymeric flocculants and suspended anionic organic compounds to generate insoluble salts and then form clots. For example, Al^{3+}, which is the main function of IA in this test, combines with OH- of anionic property in water to form a colloidal substance, which can adsorb tiny particles in the slurry, thus forming a floc substance with $Al(OH)_3$ as the core.

16.5 Conclusion

In this paper, in order to study the effect of different types of dewatering chemicals on the comprehensive effect of dewatering and curing of river and lake bottom mud, a lake in Wuhan was used as the research object, and the dewatering effect of the chemicals was analyzed by various means such as sludge specific resistance, SV30 settlement and small and medium simulated pressing tests, and their depth dewatering mechanism was elaborated, and several conclusions were drawn as follows.

(1) Different types of bottom sludge dewatering agents differ significantly in their efficiency of water removal from river and lake bottom sludge, and organic polymer dewatering agents occupy obvious advantages in dewatering efficiency and dosage.
(2) The new type of dewatering agents generated by the combination of different types of dewatering agents shows more excellent ability in dewatering performance, but their comprehensive costs are mostly higher than those of single agents, which are less suitable for project applications.
(3) The better dewatering effect of organic polymer flocculants is mainly attributed to the decrease of the bilayer potential of mud particles and the accelerated dewatering and curing of the substrate by the bridging flocculation deposition, and the difference of dewatering effect of different organic polymer flocculants is attributed to their different molecular weights.

References

Bo C (2020) Application prospect of microbial flocculants in sewage treatment. Chem Manag 30:116–117

Cao BD, Zhang T, Zhang WJ, Wang DS (2021) Enhanced technology based for sewage sludge deep dewatering: a critical review. Water Res 189:116650

Ding T, Tian YJ, Liu JB, Hu LF, Wang JY, Guo ZN, Hou J (2015) Heavy metal pollution assessment and environmental dredging depth study of river sediment in Hangzhou. J Environ Sci 35(3):911–917

Guo KY, Gao BY, Wang WY, Yue QY, Xu X (2019) Evaluation of molecular weight, chain architectures and charge densities of various lignin-based flocculants for dye wastewater treatment. Chemosphere 215:214–226

Jiang XC, Li YS, Tang XH, Jiang JY, He Q, Xiong ZK, Zheng HL (2021) Biopolymer-based flocculants: a review of recent technologies. Environ Sci Pollut Res Int 28(34):46934–46963

Kasmi A, Abriak NE, Benzerzour M, Azrar H (2017) Effect of dewatering by the addition of flocculation aid on treated river sediments for valorization in road construction. Waste Biomass Valoriz 8(3):585–597

Li M, Wang YL, Hou XB, Wan X, Xiao HN (2018) DMC-grafted cellulose as green-based flocculants for agglomerating fine kaolin particles. Green Energy Environ 3(2):138–146

Liu LX, Han YW, Liu H, Meng XJ, Gao XT, Hou CF, Xiong XY, Ke Y (2020) Dredging technology and its impact on the treatment effect of polluted water. J Environ Eng Technol 10(1):63–71

Lu T, Luo CL, Qi DM, Zhang D, Zhao HT (2019) Efficient treatment of emulsified oily wastewater by using amphipathic chitosan-based flocculant. React Funct Polym 139:133–141

Mikkelsen H, Mascarenhas T, Nielsen H (2018) Key parameters in sludge dewatering: testing for the shear sensitivity and EPS content. Water Sci Technol 46:105–114

Onoguchi A, Granata G, Haraguchi D, Hayashi H, Tokoro C (2019) Kinetics and mechanism of selenate and selenite removal in solution by green rust-sulfate. Royal Soc Open Sci 6(4):182147

Pandey A, Pathak V, Kothari R, Black PN, Tyagi VV (2019) Experimental studies on zeta potential of flocculants for harvesting of algae. J Environ Manag 231:562–569

Piló D, Carvalho AN, Pereira F, Coelho HE, Gaspar MB (2019) Evaluation of macrobenthic community responses to dredging through a multimetric approach: effective or apparent recovery. Ecol Ind 96(01):656–668

Sun YJ, Zhou SB, Sun WQ, Zhu SC, Zheng HL (2020) Flocculation activity and evaluation of chitosan-based flocculant CMCTS-g-P(AM-CA) for heavy metal removal. Sep Purif Technol 241:116737

Wei H, Gao BQ, Ren J, Li AM, Yang H (2018) Coagulation/flocculation in dewatering of sludge: a review. Water Res 143:608–631

Xu Z, Peng Y, Ya W (2022) Enhanced technology for sewage sludge advanced dewatering from an engineering practice perspective: a review. J Environ Manag 321:115938

Yang CH, Yang P, Geng J, Yin HB, Chen KN (2020) Sediment internal nutrient loading in the most polluted area of a shallow eutrophic lake (Lake Chaohu, China) and its contribution to lake eutrophication. Environ Pollut 262:114292

Zhong JC, You BS, Fan CX, Li B, Zhang L, Ding SM (2008) Influence of sediment dredging on chemical forms and release of phosphorus. Pedosphere 1:34–44

Chapter 17
Simultaneous Determination of 16 Metallic Elements in Surface Water by ICP-MS with Quadrupole Collision Cell Technology

Yin-Fu Wang and Liu Yang

Abstract In this paper, an ICP-MS with quadrupole collision cell technology (CCT) was used for the simultaneous determination of 16 metal elements in surface water samples, including Cu, Zn, Cd, Hg, Fe, Mn, Mo Co Be Sb, Ni, Ba, V, Ti, Tl, as specified in *GB3838-2002 Standard for Surface Water Environmental Quality*. The kinetic energy discrimination effect (KED mode) was applied to eliminate the mass spectrum interference. ^{6}Li ^{74}Ge ^{103}Rh ^{115}In ^{185}Re with different mass number were selected as the internal standard elements to correct the matrix effect and signal drift effectively. The method was verified under optimal instrument conditions. The verification results show that: The mass concentration of all elements had a good linear relationship with spectral intensity. The linear correlation coefficients were all greater than 0.999. The detection limits of 16 elements were 0.004–0.95 μg/L, the relative standard deviation RSDS were all less than 5.6%, and the relative errors were −4.15%–4.93%. The recoveries of surface water samples ranged from 96.3% to 106.04%. It is proved that this method has good accuracy and precision and can meet the requirements of surface water analysis and detection.

Keywords Surface water · Metallic element · ICP-MS · KED mode

With the improvement of people's health awareness and the national emphasis on ecological environment, it is particularly important to monitor and evaluate surface water bodies, and higher requirements are also put forward for the water quality of surface drinking water sources (Xin et al. 2015). To control water pollution, protect surface water quality, safeguard human health, and maintain a sound ecosystem, China issued the first edition of *the Surface Water Environmental Quality Standard* in 1983, and has since revised it three times. In the third revised version (GB3838-2002 "Surface water Environmental Quality Standard" 2002), there are 24 basic testing items, 80 specific items of surface water source for centralized drinking water and

Y.-F. Wang (✉) · L. Yang
Shenyang Center, China Geological Survey, Shenyang 110034, Liaoning, China
e-mail: 282412082@qq.com

H. Xu (ed.), *Proceedings of the 5th International Symposium on Water Resource and Environmental Management*, Environmental Science and Engineering,
https://doi.org/10.1007/978-3-031-31289-2_17

5 supplementary items, among which the limit requirements of 16 metal elements such as copper, zinc, mercury, cadmium, and lead are specified.

At present, the instruments used in the field of inorganic analysis mainly include four categories: atomic absorption spectrometer which including flame method (FLAAS) and graphite furnace method (GFAAS) (Hou et al. 2022; Xu et al. 2022), atomic fluorescence spectrometer (AFS) (Zhang et al. 2022; Liu et al. 2022a), inductively coupled plasma spectrometer (ICP-AES) (Ni et al. 2020; Lu et al. 2023) and inductively coupled plasma mass spectrometer (ICP-MS) (Liu et al. 2022b). ICP-MS has the characteristics of high sensitivity, low detection limit (up to ppt level), wide dynamic linear range (up to 109 orders of magnitude) and is widely used in the determination of trace and ultra-trace elements (Sun et al. 2020). Although ICP-MS technology is widely used in biology, environment, geology and other fields, the problem of mass spectrum interference always exists with the development of the technology. There are a variety of studies devoted to overcoming this problem. In addition to mathematical correction methods, there are also techniques to separate interfering ions and change the ionization conditions of the plasma source part (Li et al. 2010). Each technique and method have special effects, but also have limitations. In 1997, the introduction of collision cell technology (CCT) became an important breakthrough in solving the interference problem of ICP-MS. With the development of CCT, dynamic reaction cell mode (DRC) and kinetic energy discrimination mode (KED) have been increasingly applied to solve the mass spectrum interference problem (Yu 2013). The principle of KED mode is that the impact cross section of interfering ions (usually molecular ions) is larger than that of single atomic ions, so interfering ions have a greater chance of being collided and greater kinetic energy loss when flying in the collision pool. When the potential energy well set by the quadrupole is higher than the kinetic energy of interfering ions, these interfering ions cannot enter the quadrupole and are eliminated (Li et al. 2010).

In this paper, helium gas was used as collision gas, and mass spectrum interference was eliminated by KED mode. At the same time, non-mass spectrum interference was corrected by internal standard element. A method for simultaneous determination of Cu, Zn, Cd, Pb, Hg, Fe, Mn, Mo Co Be Sb, Ni, Ba, V, Ti, Tl in surface water was established.

17.1 Experiment

17.1.1 Critical Equipment and Materials

ICP-MS: ICAP-RQ Series purchase from ThermoFisher
Argon: 99.99% purity
Helium: 99.99% purity
HNO_3: GR purity
Tuning fluid: 10 ng/ml, purchase from ThermoFisher

Multielement standard solution: 100 μg/ml, get form National Analysis and Testing Center for Non-ferrous metals and electronic Materials
^{6}Li ^{74}Ge ^{103}Rh ^{115}In ^{185}Re internal standard solution: 1000 μg/ml, get from National Analysis and Testing Center for Non-ferrous metals and electronic Materials standard sample:
0.1 μg/ml multi-element mixed certified liquid standard sample GNM-M255549-2013
1 μg/ml multi-element mixed certified liquid standard sample GNM-M255548-2013
10 μg/ml multi-element mixed certified liquid standard sample GNM-M252518-2013.
Water for experiment: Deionized water, resistivity 18 MΩ·cm

17.1.2 Method

Sample handling. The treatment of water samples can be selected in the following ways according to the characteristics of water samples:

(1) Clean water samples can be directly acidified with nitric acid to pH < 2;
(2) Muddy water samples with low organic matter content should be filtered by 0.45 μm filter membrane and acidified to pH < 2.
(3) The samples with high organic matter content can be treated by microwave digestion.

The water samples in this experiment were collected from the Puhe River in Shenyang China. The samples were somewhat turbid but contained less organic matter. Only filtration and acidification were done.

Process parameter. Standard tuning solution was used to optimize the analytical parameters of the instrument. Preheated the instrument for 30 min. Sensitivity, oxide and double charge of the instrument are tuned with the mass spectrum tuning solution. When all three parameters meet the requirements, the quality correction and resolution verification of the instrument were carried out to obtain the optimal working parameters of the instrument, as follows: radio frequency power 1250 W; Atomizer flow rate 0.82 L/min; The flow rate of the auxiliary device is 0.7 L/min; The flow rate of cooling gas is 13 L/min, and the flow rate of collision gas (helium) is 3.2 ml/min. Integration time 0.5 s; Semiconductor cooling temperature 2–4 °C; Scanning mode: peak jumping; Number of scans: 100.

17.2 Results and Discussion

17.2.1 Establishment of Experiment Methods

Under the best working condition of the instrument, the establishment of ICP-MS mass spectrometry includes the selection of the isotope of elements to be measured, the internal standard elements and the determination mode of the instrument. The selection of element isotopes generally follows the principle of large abundance and small interference. The selection of internal standard elements should follow the principle that there is no internal standard element in the sample, the internal standard element is not interfered by the sample matrix or there is no spectral line interference between the internal standard element and the target element, the mass number of the internal standard element and the element to be measured is as close as possible, and the internal standard element and the element to be measured have similar ionization potential, so as to ensure that the behavior and performance of the two elements are basically the same under the same conditions. The instrument operating mode of ICP-MS can be divided into standard mode (STD) and KED mode according to the operating principle. The standard model is applicable to the analysis of elements with little or no mass spectrum interference, such as Be, Hg, Tl, etc. According to the characteristics of quality spectrum interference of the element to be measured, different working modes are selected to analyze and test. The specific test parameters are shown in Table 17.1.

17.2.2 Optimization of Interference Problems for ICP-MS

The interference problems of ICP-MS mainly include non-mass spectrum interference and mass spectrum interference. Non-mass spectrum interference mainly consists of matrix interference and physical effect interference. Non-mass spectrum interference is represented by signal suppression and signal drift. The most effective method to eliminate matrix effect in ICP-MS analysis is internal standard correction, which can compensate matrix effect obviously. During the long-term analysis of ICP-MS, the oxides and matrix particles of the elements to be measured will be deposited in the cone mouth of the sampling cone and the intercepting cone, resulting in a smaller aperture, which will affect the stability of the analysis signal and produce obvious physical effect interference. The addition of internal standard elements can effectively monitor and correct the short- and long-term signal drift of the instrument. According to the selection principle of internal standard element, ^{6}Li ^{74}Ge ^{103}Rh ^{115}In ^{185}Re is selected as the internal standard element in this paper.

The interference of mass spectrum is mainly polyatomic ion interference and isotopic interference. Multi - atomic ion interference includes argon—based ion and oxide ion interference. The interference of argon-based ions comes from working

Table 17.1 ICP-MS test parameters

Elemental isotope	Abundance (%)	Major interference	Internal standard element	Mode	Elemental isotope	Abundance (%)	Major interference	Internal standard element	Mode
^{63}Cu	69.17	$^{40}Ar^{23}Na^+$	Ge	KED	^{9}Be	100		Li	STD
^{66}Zn	65.93	$^{14}N^{52}Cr^+$	Ge	KED	^{121}Sb	57.3	$^{40}Ar^{81}Br^+$	In	KED
^{74}Cd	36.5	$^{40}Ar^{34}S^+$	Ge	KED	^{58}Ni	68.2	$^{12}C^{46}Ti$	Ge	KED
^{106}Pb	27.33	$^{16}O^{1}H^{+89}Y$	Rh	KED	^{138}Ba	71.7	$^{121}Sb^{16}O^{1}H^+$	In	KED
^{56}Fe	91.72	$^{40}Ar^{16}O^+$	Ge	KED	^{51} V	99.75	$^{35}Cl^{16}O^+$	Ge	KED
^{55}Mn	100	$^{40}Ar^{14}N^{1}H^+$	Ge	KED	^{48}Ti	73.8	$^{32}S^{16}O^+$	Ge	KED
^{98}Mo	24.13	$^{81}Br^{16}O^{1}H^+$	Rh	KED	^{205}Tl	70.48		Re	STD
^{59}Co	100	$^{14}N^{45}Sc$	Li	KED	^{202}Hg	29.8		Re	STD

gas and solvent which can be suppressed by different working modes. The interference sources of oxide ions are solvents and environment which can be inhibited by tuning the instrument to reduce the yield of oxide products. The interference of the same amount of ectopic element mainly refers to the interference ions with the same atomic mass as the target element. To eliminate the interference of the same amount of ectopic element, other isotopes can be selected to avoid the interference. For the interference of single isotope element, interference correction formula can be adopted to deduct. Thermo Fisher's ICAP-RQ series instruments used in this paper have QCell technology combined with Flatapole's low mass number elimination function and helium KED anti-jamming technology, which can solve the mass spectrum interference problem in the test process.

17.2.3 Standard Work Curve

Mixed standard solutions with mass concentrations of 0 μg/L, 10 μg/L, 20 μg/L, 50 μg/L, 100 μg/L, 0.2 mg/L, 0.5 mg/L, 1 mg/L and 2 mg/L were prepared using 1% nitric acid solution. With the mass concentration of the measured results as the horizontal coordinate and the intensity as the vertical coordinate, the standard curve of each element and correlation coefficient are automatically drawn by the software. As can be seen from Table 17.2, the standard curves of all elements to be tested have a good linear relationship, with linear correlation coefficients greater than 0.999.

17.2.4 Method Detection Limits

The detection limit of elements was repeated 7 times according to the blank solution of high purity nitric acid (1 + 99). The detection limit of the method was calculated according to the formula ($MDL = t_{(n-1,0.99)} \times S$). The calculation results are shown in Table 17.3. The detection limit of elements to be tested is between 0.004–0.95 μg/L, which is far lower than the minimum detection limit requirements in GB3838-2002 "Surface Water Environmental Quality Standards".

17.2.5 The Precision and Accuracy of the Method

Three certified standard samples with different mass concentrations were used to test the precision and accuracy of the method. The test results are shown in Table 17.4. The relative standard deviation of 16 elements is less than 5.6%, and the relative error is −4.15%–4.93%, indicating that the method has good precision and accuracy.

Table 17.2 Standard curve and correlation coefficient

Element	Linearity range μg/L	Equation of linear regression	Correlation coefficient
Cu	0–2000	y = 0.981x + 1.0253	0.9997
Zn	0–2000	y = 0.9803x + 1.1755	0.9993
Cd	0–500	y = 0.9889x–0.099	0.9997
Pb	0–500	y = 1.056x + 0.035	0.9997
Fe	0–2000	y = 0.9969x + 0.0041	0.9999
Mn	0–2000	y = 0.996x + 0.3018	0.9999
Mo	0–1000	y = 0.9972x + 0.2101	0.9999
Co	0–500	y = 0.9946x + 0.4138	0.9997
Be	0–500	y = 1.0998x + 0.1251	0.9999
Sb	0–500	y = 1.0359x–0.012	0.9997
Ni	0–500	y = 1.0162x + 0.0187	0.9999
Ba	0–500	y = 0.9925x + 0.5735	0.9995
V	0–500	y = 1.0024x–0.1794	1.0000
Ti	0–500	y = 0.9987x + 0.1022	0.9999
Tl	0–500	y = 0.9904x + 0.6938	0.9995
Hg	0–500	y = 0.9814x + 0.2985	0.9999

Table 17.3 Detection limits (μg/L)

Element	Detection limit	Minimum detection limits	Element	Detection limit	Minimum detection limits
Cu	0.10	1.00	Be	0.07	0.02
Zn	0.70	50.00	Sb	0.21	0.25
Cd	0.03	1.00	Ni	0.08	2.48
Pb	0.09	10.00	Ba	0.25	6.18
Fe	0.95	30.00	V	0.11	6.98
Mn	0.14	10.00	Ti	0.49	0.40
Mo	0.06	2.30	Tl	0.004	0.004
Co	0.04	1.90	Hg	0.014	0.05

17.2.6 Analysis of Surface Water Samples

The surface water samples collected from the Puhe River in Shenyang China were treated by adding standard solutions of different mass concentrations of elements to be measured. The test method recommended in *GB3838-2002 "Surface Water Environmental Quality Standards*" and new method were used to measure the sample with adding standard recovery, and the recovery rate was calculated according to the test results of the water sample and the spiked water sample. As can be seen from

Table 17.4 Precision and accuracy

Sample No	Element	Mean value μg/ml	SD	RSD (%)	Sample concentration μg/ml	RE (%)
M255549	Cu	0.098	0.002	2.53	0.1	−1.81
	Zn	0.101	0.003	2.53	0.1	0.95
	Cd	0.091	0.002	1.66	0.1	−4.15
	Pb	0.100	0.000	0.41	0.1	0.15
	Fe	0.099	0.002	1.73	0.1	−1.47
	Mn	0.098	0.002	1.62	0.1	−2.47
	Mo	0.099	0.001	1.45	0.1	−1.44
	Co	0.097	0.001	1.02	0.1	−3.37
	Be	0.103	0.002	1.56	0.1	3.47
	Sb	0.103	0.001	0.79	0.1	2.57
	Ni	0.104	0.001	0.71	0.1	3.98
	Ba	0.102	0.002	2.05	0.1	1.85
	V	0.103	0.002	1.61	0.1	3.09
	Ti	0.103	0.002	1.58	0.1	2.54
	Tl	0.103	0.002	2.31	0.1	3.41
	Hg	0.100	0.000	0.33	0.1	0.37
M255548	Cu	1.072	0.053	4.93	1	4.16
	Zn	0.984	0.016	1.62	1	−1.57
	Cd	0.989	0.022	2.27	1	−1.13
	Pb	1.021	0.057	5.56	1	2.10
	Fe	1.110	0.029	2.59	1	4.02
	Mn	0.997	0.053	5.33	1	−0.32
	Mo	1.074	0.056	5.22	1	4.41
	Co	1.072	0.056	5.22	1	4.17
	Be	0.996	0.026	2.59	1	−0.36
	Sb	1.066	0.036	3.35	1	3.59
	Ni	0.997	0.040	4.01	1	−0.32
	Ba	1.058	0.029	2.74	1	2.83
	V	1.014	0.024	2.39	1	1.39
	Ti	1.028	0.018	1.73	1	2.81
	Tl	1.010	0.024	2.41	1	0.98
	Hg	1.042	0.038	3.61	1	4.15
M252518	Cu	10.079	0.160	1.59	10	0.79
	Zn	11.048	0.222	2.01	10	0.48
	Cd	9.725	0.091	0.94	10	−2.75

(continued)

Table 17.4 (continued)

Sample No	Element	Mean value μg/ml	SD	RSD (%)	Sample concentration μg/ml	RE (%)
	Pb	9.889	0.168	1.70	10	−1.11
	Fe	10.570	0.107	1.02	10	2.70
	Mn	10.463	0.344	3.29	10	4.63
	Mo	10.095	0.175	1.74	10	0.95
	Co	10.529	0.449	4.26	10	2.29
	Be	9.644	0.414	4.30	10	−3.56
	Sb	10.384	0.103	0.99	10	3.84
	Ni	10.493	0.221	2.11	10	4.93
	Ba	10.485	0.152	1.45	10	4.85
	V	10.116	0.174	1.72	10	1.16
	Ti	10.023	0.127	1.26	10	0.23
	Tl	10.073	0.117	1.17	10	0.73
	Hg	10.660	0.095	0.89	10	1.60

Table 17.5, the recoveries of the spiked water samples with the recommended method ranged from 95.44% to 106.26%, while that of the spiked water samples with new method ranged from 96.3% to 106.04%. There was little difference between the recoveries of the two analytical methods.

17.3 Conclusion

This paper verifies the feasibility of simultaneous determination of 16 metal elements in surface water by ICP-MS with quadrupole collision cell technology. The experimental results show that the detection limit of elements to be measured is far lower than the minimum detection limit requirements of GB3838-2002 "Surface Water Environmental Quality Standards", and the mass concentration of all elements to be measured has a good linear relationship with the spectral intensity. The precision and accuracy of the analytical method and the recovery rate of the sample can meet the quality requirements of water sample analysis in DZ/T0130-2006 Standard for Quality Management of Geological and Mineral Laboratory Testing. The test method recommended in GB3838-2002 standard and the new ICP-MS method were used to carry out the recovery test of Shenyang Puhe surface water samples respectively. The recovery rates of the two different analysis methods were basically no difference, which further verified the accuracy of the test results of the new method. And the new method has great advantages of the linear range, testing efficiency and chemical reagent consumption. The application of ICP-MS with quadrupole CCT in the determination of surface water samples can effectively solve the problem of mass

Table 17.5 Actual sample test

Element	Selected standard analytical methods			Method of this paper
	Method	Serial number	Recovery	Recovery
Cu	Atomic absorption and chelation extraction	GB7475-87	97.79	102.05
Zn	Atomic absorption—direct method	GB7475-87	100.83	96.91
Cd	Atomic absorption and chelation extraction	GB7475-87	99.86	100.49
Pb	Atomic absorption and chelation extraction	GB7475-87	95.44	99.23
Fe	Flame atomic absorption spectroscopy method	GB11911-89	102.60	106.04
Mn	Flame atomic absorption spectroscopy method	GB11911-89	97.03	105.79
Mo	Flameless atomic absorption method	GB/T5750-2006	104.16	97.11
Co	Flameless atomic absorption method	GB/T5750-2006	98.58	98.30
Be	Graphite furnace atomic absorption spectrometry	HJ/T59-2000	104.85	99.11
Sb	Hydrogenated atomic absorption spectrometry	GB/T5750-2006	104.77	99.76
Ni	Flameless atomic absorption method	GB/T5750-2006	105.73	96.30
Ba	Flameless atomic absorption method	GB/T5750-2006	102.01	103.73
V	Flameless atomic absorption method	GB/T5750-2006	102.19	102.38
Ti	Salicyl-fluoroketone spectrophotometry	GB/T5750-2006	99.83	102.46
Tl	Flameless atomic absorption method	GB/T5750-2006	106.26	98.58
Hg	Cold-vapor atomic absorption method	GB7468-87	100.54	102.92

spectrum interference in sample testing, greatly improve the efficiency of sample analysis, and reduce the environmental pollution caused using chemical reagents and the health hazards of the analysis and testing personnel.

References

GB3838–2002 "Surface water Environmental Quality Standard" (2002)

Hou YR, Lu JL, Fan YC (2022) Uncertainty evaluation and method improvement of determination of copper, lead, and zinc in rocks by atomic absorption spectrometry. Spectrosc Spectr Anal 42(7):2101–2106

Li B, Hu JY, Zhao MT (2010) The Performance and application on the collision/reaction cell ICP-MS. J Chin Mass Spectrom Soc 31(1):1–11

Liu P, Du MF, Li ZY (2022a) Determination of trace tellurium content in steel by hydride generation atomic fluorescence spectrometry. Spectrosc Spectr Anal 42(10):3103–3108

Liu XL, Sun WJ, Ren YZ (2022b) Determination of sliver, tin and boron in soil or sediment samples with microwave digestion by mixed mode inductively coupled plasma-mass spectrometry. J Chin Mass Spectrom Soc 43(4):522–531

Lu K, Li YQ, Ni XL (2023) Determination of eighteen elements in soil by inductively coupled plasma optical emission spectrometry with super microwave digestion. Chin J Inorg Anal Chem 13(2):123–128

Ni ZL, Han SF, Mo RH (2020) Determination of nine metal elements in forest soil by inductively coupled plasma atomic emission spectrometry after acid-digestion. J Anal Sci 36(2):304–308

Sun LL, Song JM, Liu Y (2020) Simultaneous determination of molybdenum and other heavy metals in Mytilus edulis by inductively coupled plasma mass spectrometry with quadrupole collision cell technology. Mar Environ Sci 39(3):453–459

Xin XD, Hu F, Sun L (2015) Determination of 10 kinds of metal elements in water by inductively coupled plasma-mass spectrometry based on kinetic energy discrimination. Chem Anal Meterage 24(6):35–38

Xu XH, Zhang WY, Gu ZL (2022) Determination of Cd, Pb in soil by graphite furnace atomic absorption spectrometry. Chin J Inorg Anal Chem 12(3):12–16

Yu X (2013) The present situation and advance in collision/reaction cell technique of inductively coupled plasma quadrupole mass spectrometry. Metallurg Anal 33(3):14–23

Zhang Q, Wu J, Zheng Y (2022) Determination of mercury, arsenic, selenium and antimony in soil by atomic fluorescence spectrometry with graphite digestion. Phys Test Chem Anal Part B: Chem Anal 58(6):664–669

Chapter 18
Spatio-Temporal Variation Characteristics of NPP in Weihe Watershed and Its Response to Environmental Factors Based on the CASA and CA–Markov Model

Lixia Wang, Feiyan Pan, Mingshuang Zhang, Zhao Liu, Shuangcheng Zhang, and Jinling Kong

Abstract In this study, in order to simulate and predict NPP in the Weihe Watershed and examine its spatiotemporal coverage and dynamic variations, CASA and CA-Markov models are linked. To statistically assess the NPP response to environmental conditions, correlation analysis was used. Results revealed: (1) Seasonal fluctuations were noticeable on the time scale within a year, with the NPP in July being the highest and the NPP in January being the lowest. The forecast indicates that the vegetation of the watershed will continue to increase over the coming ten years based on the rising trend in the interannual change. (2) The NPP coverage is noticeably varied on a regional scale, with a generally high coverage in the south and east and a low coverage in the north and west. (3) The response of NPP to environmental conditions is notable but diverse. There was a weakly positive correlation between the response and climate parameters. With increasing slope and altitude, the NPP exhibits a bimodal tendency that continues to rise. The northern and western aspects' NPP levels, meanwhile, were higher. (4) The high coupling degrees of the CASA and CA-Markov models are appropriate for the prediction of NPP in the Watershed.

Keywords Net primary productivity (NPP) · Simulation and prediction · Environmental factor response · Coupling model · Weihe Watershed

---

L. Wang · F. Pan (✉) · S. Zhang · J. Kong
School of Geological Engineering and Geomatics, Chang'an University, Xi'an 710054, China
e-mail: 2022226043@chd.edu.cn

M. Zhang
School of Earth Sciences and Resources, Chang'an University, Xi'an 710054, China

Z. Liu
School of Environmental Science and Engineering, Chang'an University, Xi'an 710054, China

H. Xu (ed.), *Proceedings of the 5th International Symposium on Water Resource and Environmental Management*, Environmental Science and Engineering,
https://doi.org/10.1007/978-3-031-31289-2_18

18.1 Introduction

The 19th National Congress of the Communist Party of China (CPC) put forward the slogan of "treating the ecological environment like life", and under the guidance of this concept, we will move towards a new era of vegetation and ecological civilization. Net Primary Production (NPP) is the net increase in the amount of energy fixed by photosynthesis in a given time (Guan et al. 2017), and is a measure of the productivity of vegetation communities under natural conditions (Zheng et al. 2020). It is also the study basis of the movement of matter and energy in terrestrial ecosystems, which represents important variables of vegetation activities and the main factors that regulate ecological processes. Moreover, it plays a sinister role in analysing climate change (Naeem et al. 1985).

Since the 1970s, the research on vegetation NPP has gained increasing attention. Many relevant calculation models have emerged, including statistical, process, and parameter models (Comparison Between Modeled et al. 2002; Bian et al. 2010; Bradford et al. 2005; Cao et al. 2004; Defosse and Bertiller 1991; Field et al. 1995). NPP estimation models extend site data to regional and even the whole world; the Carnegie-Ames-Stanford Approach (CASA) model is a classical representative of this model (Ruimy et al. 1994; Kalubarme et al. 2003; Lewis et al. 1998; Jones et al. 2003; Prasad et al. 2006; Nemani et al. 2003; Doraiswamy et al. 2003). In particular, after the optimization of this model by Zhu et al. in 2007, its applicability was improved and it could be applied well in regional spaces of different scales (Zhu et al. 2007).

Many studies have been performed by domestic and international scholars on scientific issues such as the spatial and temporal variability of NPP dynamics and the driving factors, and more fruitful results have been achieved (Li et al. 2001; Lai et al. 1982; Shang et al. 1960). For example, Nemani et al. studied the global climate change for about 20 years at the end of the last century and concluded that it was favorable for vegetation growth (Nemani et al. 2003). Bokhorst et al. estimated the vegetation restoration after extreme winter warming events in subarctic regions. Neumann et al. studied the difference between vegetation NPP products from Moderate Resolution Imaging Spectroradiometer (MODIS) and the inventory data of Austria's terrestrial national forest (Neumann et al. 2016). Zhu et al. systematically analyzed the NPP and corresponding climatic data of China's terrestrial vegetation from 1982 to 1999 (Zhu et al. 2007). Chen et al. studied the effects of climate factors on vegetation NPP in different periods in China (Chen et al. 2011). Meanwhile, scholars have also conducted a lot of studies on the Spatio-temporal characteristics of vegetation NPP and its response to climate change in several regions of China, such as the Qinghai-Tibet Plateau, Loess Plateau, northeastern region, southeastern region, and subtropical mountainous hilly region, etc. These results are important for exploring the spatial and temporal patterns, change processes and driving mechanisms of terrestrial ecosystem NPP (Yu et al. 2018; Neumann et al. 2015; Wang et al. 2019; Kong et al. 2019; Zhang et al. 2019; Luo et al. 2018). In addition, remote sensing technology can provide important information for the dynamic change process of vegetation type/cover. The combination of remote sensing model and mathematical model can realize dynamic monitoring and simulation prediction of vegetation resources, so as to carry out the disaster risk assessment and crop yield estimation. For example, while the effect of precipitation on NDVI has a lag, Funkc et al. developed an empirical model based on this relationship and predicted NDVI changes in semi-arid regions of Africa (Funk and Brown 2006). Goncalves et al. used multivariable and univariate models to predict the crop NDVI in the Sao Paulo sugarcane producing area in Brazil (Gonçalves et al. 2012).

The Weihe Watersh (33°–38° N, 104°–110° E) is located in the arid and semi-arid region of the northwestern interior, with a total area of about 134,766 km^2 (Fig. 18.1) (Ji and Peters 2004). The main vegetation types include cultivated vegetation, broad-leaved forest and grassland, etc. The tributaries of the Jinghe and Luohe rivers on the north bank flow through the Loess Plateau and carry a large amount of sediment. The ecological and environmental problems such as v including the deterioration of the vegetation, the desertification of the land, the loss of the soil and the seasonal change of the river are very prominent (Wang et al. 2018; Zhang and Zhang 2008). At the same time, the Guantian Economic Zone in the watershed is a strategically important place connecting the east and the west, the north and the south. It is an area with a good economic foundation, profound humanistic history, and great development potential in western China. It is also an important fulcrum for realizing the construction and development of the country's "The Belt and Road". Therefore, studying the NPP of the watershed and its response to environmental drivers is of great significance for understanding the carbon cycle characteristics of regional terrestrial ecosystems, rational utilization of vegetation resources, evaluation of the benefits of ecological construction projects, and formulation of watershed ecological construction and sustainable development strategies.

Our study used MODIS NDVI products (MOD13), meteorological data, vegetation type data, and DEM data, etc. In the context of comprehensive consideration of the effect time of the Returning Farmland to Forest Program, Hanjiang to Weihe Watershed Project, by coupling CASA model and CA–Markov model, firstly, the Spatio-temporal coverage of NPP in the Weihe Watershed in 2000, 2006, 2012, and 2018 was calculated, followed by quantitative analysis of its response to precipitation, temperature, elevation, slope, and slope orientation, and finally, the spatial coverage pattern of NPP in the Weihe Watershed in 2024 and 2030 was predicted.

18.2 Materials and Methods

18.2.1 Data Sources and Preprocessing

Remote sensing data: The data selected in this study are MOD13Q1 data from the NASA website (https://ladsweb.modaps.eosdis.nasa.gov/). The spatial resolution was 250 m. To facilitate the reading and processing of ENVI and ArcGIS software as

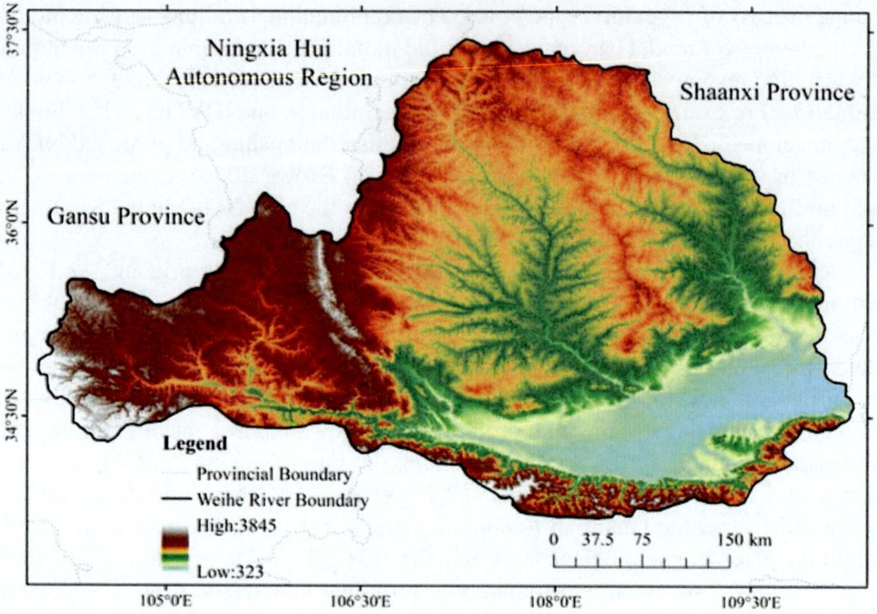

Fig. 18.1 Geographical location and topography of Weihe Watershed

well as the spatial matching of the existing map data, preprocessing such as splicing, projection conversion, and format conversion is required. Therefore, Modis Reprojection Tools (MRT) were used for preprocessing. At the same time, the image data is affected by the atmosphere, soil, height angle, and other factors, resulting in noise in the NDVI value. Therefore, the maximum value synthesis method is selected to synthesize the two phases of NDVI data every month.

Meteorological data: The data mainly include sunshine hours, temperature and precipitation, which were obtained from the National Meteorological Information Center (http://data.cma.cn/), including 33 meteorological stations in and around the Weihe Watershed. As there are few stations observing radiation data nationwide, the radiation derivation formula is used to convert sunshine hours into radiation data. Finally, sunshine radiation, average daily temperature, and average daily precipitation data were interpolated into raster data with spatial resolution consistent with NDVI using the Kriging interpolation method.

Other data: Digital elevation model (DEM) data, administrative boundary data, vegetation type data, etc., were derived from the geospatial data cloud (http://www.gscloud.cn/). DEM data were spliced and clipped in ArcGIS to obtain DEM data covering the study area. Then, the spatial resolution was resampled to 250 m and slope and aspect information was extracted for overlay analysis with NPP data.

18.2.2 Method

CASA model. The CASA model is a parametric model based on the light energy utilization principle developed by Potter and Field according to Monteith's theory of calculating NPP using photosynthetically active radiation (APAR) and efficiency for solar energy utilization (ε) (Field et al. 1995, 1998; Potter et al. 1993; Zhu et al. 2006). The model has been widely used in large-scale vegetation NPP simulations and global vegetation carbon cycles, and has been calibrated by more than 1,900 field sites worldwide (Zhu et al. 2004). In this study, the improved CASA model of Zhu Wenquan was used to estimate the vegetation NPP of terrestrial ecosystems in the Weihe Watershed in 2000, 2006, 2012, and 2018.

Based on the physiological and ecological processes of vegetation, the model-simulated NPP depends on the photosynthetically active radiation absorbed by vegetation and efficiency of solar energy utilization (Zhu et al. 2005). The calculation formula is as follows:

$$NPP(x,t) = APAR(x,t) \times \varepsilon(x,t)$$

where t is the time and x is the position in space.

Correlation analysis. The detailed information about this method can be achieved from reference (Wang et al. 2018).

CA–Markov model. The CA–Markov model integrates Cellular Automata (CA) with spatial evolution capability and Markov chain with temporal reasoning function, which plays a significant role in studying the spatiotemporal evolution law and characteristics of complex spatial systems. The model uses the transition matrix and suitability atlas between the two periods to redistribute the base period data, so as to simulate the future spatial coverage characteristics of the data. In this paper, the CA-Markov module in the DRISI17.0 platform is used to predict the future spatial coverage pattern of net primary productivity of vegetation.

18.3 Results and Analysis

18.3.1 Temporal Variation Characteristics of the NPP

It can be seen that the intra-year variation of NPP shows a unimodal curve, with the highest value of the year occurring in July, about 115.29 gC/m^{2}a, and the lowest value in January, about 10.75 gC/m^{2}a. The seasonal variation of NPP showed an obvious periodicity, which was basically consistent with the variation of vegetation phenological characteristics. Vegetation in summer is in a vigorous growth stage, with the highest NPP value, about 382.08 gC/m^{2}a; in spring temperature warms up, precipitation increases, and vegetation growth speeds up, so NPP is second only to summer, about 129.38 gC/m^{2}a; in autumn vegetation gradually stops growing until withered, the NPP value gradually decreased, slightly lower than that in spring, about 118.94 gC/m^{2}a; the NPP of winter vegetation was the lowest, about 42.97 gC/m^{2}a (Fig. 18.2).

The inter-annual change of NPP shows an overall upward trend. From 2000 to 2006, there was a significant increase, and the average annual NPP values were about 549.12 and 700.88 gC/m^{2}a, respectively. The NPP value in 2012 was basically the same as in 2006, about 770.85 gC/m^{2}a. The NPP value in 2018 increased slightly, about 828.00 gC/m^{2}a. It shows that the vegetation growth in the Weihe Watershed has a trend toward positive development (Fig. 18.2).

18.3.2 Spatial Coverage Characteristics of the NPP

The spatial coverage of NPP is similar from year to year, basically showing the characteristics of high north, low east and low west, with obvious spatial differences, as shown in Fig. 18.4. Among them, the Qinling and the area above Zhuangtou on the Beiluohe River areas have higher NPP annual average values. The main reason is that most of these areas are warm temperate zone, vegetation is evergreen broad-leaved forest or deciduous broad-leaved forest, and the combination of water and heat is better; the annual average value of NPP in the Guanzhong Plain, the area above Zhangjiashan on the Jinghe River and the area above Baojixia on the Weihe River is relatively small. The main reason is that the vegetation types in these areas are mainly low cultivated plants, and as the latitude rises, the temperature and precipitation gradually decreases; the annual average value of NPP in the northern part of the Weihe Watershed is the lowest because of the poor water and heat conditions due to the combined effects of latitude and topography (Figs. 18.3 and 18.4).

Fig. 18.2 NPP change trend in different months/seasons/years in the Weihe Watershed

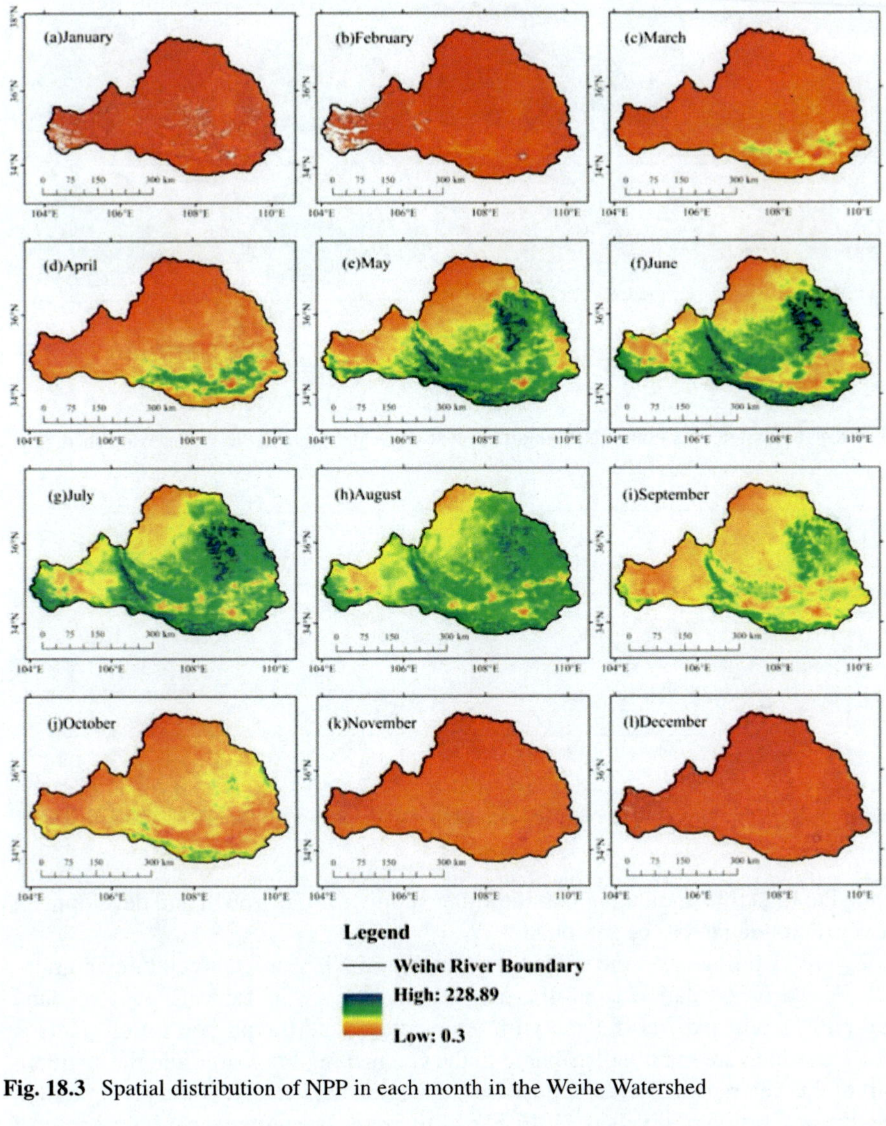

Fig. 18.3 Spatial distribution of NPP in each month in the Weihe Watershed

18.3.3 Response of NPP to Major Climatic Factors

Figure 18.5 shows the general relationship between average monthly NPP, temperature, and precipitation. As can be seen, NPP has significant exponential correlation and power function correlation with temperature and precipitation, respectively. NPP increases with the increase in temperature. When the monthly average temperature is above 20 °C, NPP tends to reach a maximum. NPP also increases with the increase in precipitation. When the monthly average precipitation exceeded 100 mm, NPP

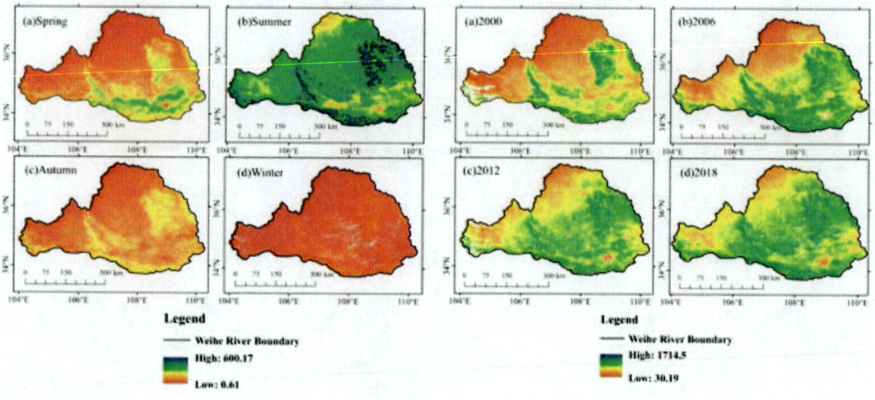

Fig. 18.4 Spatial distribution of NPP in different seasons and years in the Weihe Watershed

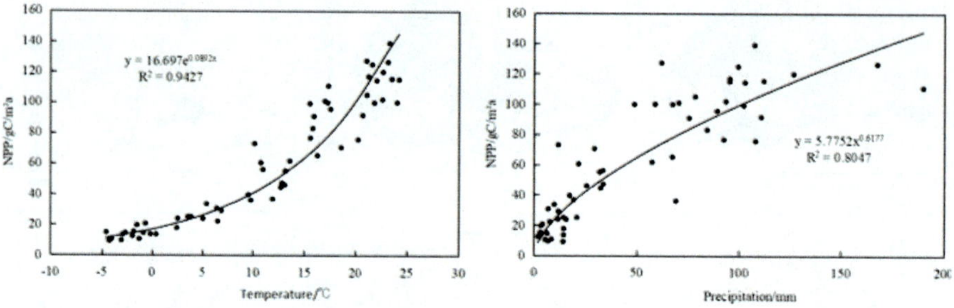

Fig. 18.5 Relationship between NPP and temperature/precipitation in the Weihe Watershed

tended to be stable, indicating that the water supply for the growth and development of vegetation tended to be saturated.

Figure 18.6 illustrates the spatial relationship of NPP with temperature and rainfall. For each correlation level, the pixel count is given in Table 18.1. It is found that NPP is positively correlated with temperature, and the proportion of pixels is 77.87%, which are mainly distributed in the Guanzhong Plain area and the southern part of the watershed. There is a positive correlation between NPP and precipitation, and the proportion of pixels is 71.46%, mainly in the western and central regions of the watershed. In general, the NPP of the Weihe Watershed has a higher degree of response to temperature.

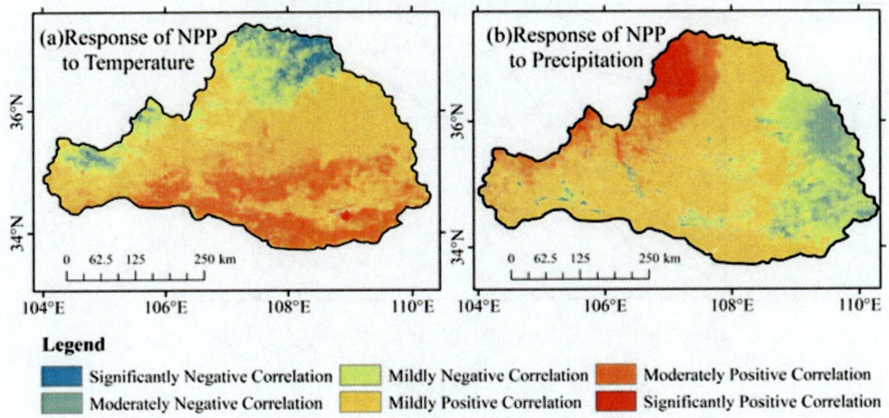

Fig. 18.6 Spatial correlation between NPP, temperature, and precipitation in the Weihe Watershed

Table 18.1 Correlation statistics of NPP, temperature and precipitation in the Weihe Watershed

Correlation level	Temperature		Precipitation	
	Number of pixels	Proportion/%	Number of pixels	Proportion/%
Significantly negative correlation	35,643	1.65	20,014	0.93
Moderately negative correlation	103,959	4.82	165,093	7.66
Mildly negative correlation	337,794	15.67	430,043	19.94
Mildly positive correlation	1,205,939	55.93	1,249,134	57.93
Moderately positive correlation	458,977	21.29	212,674	9.86
Significantly positive correlation	13,926	0.65	79,197	3.67

18.3.4 Response of NPP to Topographic Factors

Response of NPP to elevation. Figure 18.7 shows the variation of vegetation NPP with elevation. It can be seen that the average value of vegetation NPP shows a bimodal trend with increasing altitude, and there is a large gap between the maximum value and the minimum value. In the area with an elevation of less than 1250 m, namely, the Guanzhong plain area, the corresponding mean NPP fluctuates less, remaining between 732.31 and 800.94 gC/m^{2}a, which accounts for 35.77% of the total number of pixels. In the range of elevation from 1250 to 1750 m, the mean vegetation NPP decreased slightly, accounting for the highest percentage of the total number of pixels (43.75%); it was mainly distributed in the Jinghe and Beiluohe

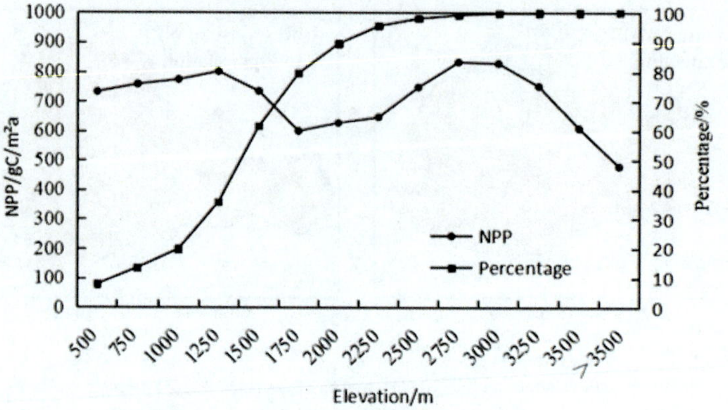

Fig. 18.7 Changes of NPP with elevation in the Weihe Watershed

Watershed. As the elevation rose to 2750 m, the mean vegetation NPP reached the maximum value, which is mainly caused by the high vegetation coverage and mostly deciduous broad-leaved forests distributed in the Qinling Mountains. The percentage of the total number of pixels decreased slightly, to 19.71%. After that, as the elevation continued to rise, the vegetation NPP gradually decreased.

Response of NPP to slope. The slope of the Weihe watershed is between 0° and 73°; the slope of more than 90% of the region is less than 30°. Figure 18.8 shows the variation of the mean NPP with the increase in slope. In general, with the increase in slope, the average NPP of the vegetation shows an increasing trend. Areas with a gentle terrain with a slope of less than 35° have fertile soil, sufficient water, and good natural conditions, which are more suitable for the growth and development of vegetation. The vegetation types in the 0–35° slope range are grasses, shrubs, cultivated plants and other short vegetation that are easily affected by the slope. When the slope exceeds 35°, the soil conditions change, and the vegetation types gradually change to coniferous forests and broad-leaved forests. Leaf forests and other vegetation that are not easily affected by slope, the depth, width, and direction of valleys begin to restrict the growth of dwarf vegetation. As coniferous forests and broad-leaved forests occupy a relatively small area in the Weihe Watershed, and grasslands and shrubs occupy a large area, the maximum value begins to decrease.

Response of NPP to aspects. It can be seen that the mean NPP is between 688.3 and 732.34 gC/m^2a. However, there are some differences between the aspects. The mean value of vegetation NPP is as follows from large to small: northwest slope > north slope > northeast slope > west slope > southeast slope > east slope > south slope > southwest slope > flat land. Differences in hydrothermal conditions can explain this feature. According to the statistics of the average temperature and precipitation in each aspect, the difference in the average annual precipitation in each aspect is approximately 1–3 mm. Meanwhile, the average temperature of the north slope and the west slope is approximately 0.05–0.20 °C lower than that of the southern and

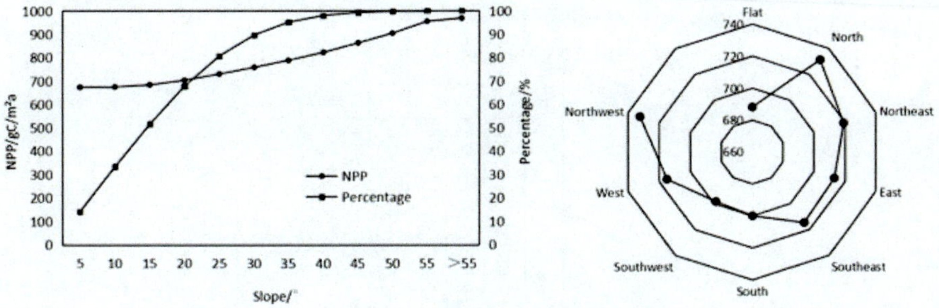

Fig. 18.8 Changes of NPP with slope/aspect in the Weihe Watershed

eastern slopes. This indicates that in northwest China, the climate is dry and sunny, the north slope is on the dark side, the solar radiation received is relatively less, the transpiration is weak, and the water is easy to retain. Therefore, the northern slope is relatively humid and the respiration is weakened, which is more conducive to the growth of vegetation. There are many results showing that the soil of the north slope and west slope has better water holding capacity, higher soil moisture content. Therefore, it is conducive to the growth of vegetation. than that of the southern and eastern slopes.

18.3.5 Prediction of the NPP Spatial Distribution Pattern

As the input data in the CA–Markov model needs to be spatial and state-discrete raster data, according to the natural discontinuity point classification method, they are divided into five levels. The area and percentage of each NPP grade in 2012 and 2018 were calculated as shown in Table 18.2. The overall characteristics of the NPP status in the Weihe Watershed in 2012 mainly correspond to lower-value regions, medium-value regions, and higher-value regions, which account for 74.74% of the total area. By 2018, the spatial coverage of very low-value regions, higher-value regions, and very high-value regions shrinked, while those of lower-value regions and medium-value regions expanded.

To verify the feasibility of the CA–Markov model in predicting the dynamic change of vegetation in the Weihe Watershed, this research took the NPP spatial coverage of 2006 and 2012 as basic data, simulated the NPP spatial coverage in 2018, and compare with the real value. The Kappa coefficient was calculated as 0.8776. It shows that the simulation results meet the accuracy requirements, and the simulation method and process have good applicability.

Therefore, based on the spatial coverage of NPP in the Weihe Watershed in 2012 and 2018, the CA_Markov model was established to predict the spatial coverage of NPP in 2024 and 2030. The results are shown in Fig. 18.9 and Table 18.2.

Table 18.2 NPP statistics each grade in the Weihe Watershed in 2012, 2018, 2024 and 2030

NPP level	2012		2018		2024		2030	
	Area/km^2	Proportion/%	Area/km^2	Proportion/%	Area/km^2	Proportion/%	Area/km^2	Proportion/%
Very low-value regions	16,229.31	12.33	13,856.88	10.53	5,364.56	4.07	2,688.94	2.04
Lower-value regions	27,095.44	20.58	31,290.50	23.77	18,493.56	14.05	12,577.31	9.55
Medium-value regions	35,644.81	27.08	38,787.81	29.46	35,145.38	26.7	32,357.50	24.58
Higher-value regions	36,102.51	27.42	31,773.88	24.14	44,656.69	33.92	49,612.63	37.69
Very high-value regions	16,569.31	12.59	15,932.31	12.1	27,981.19	21.26	34,405.00	26.14

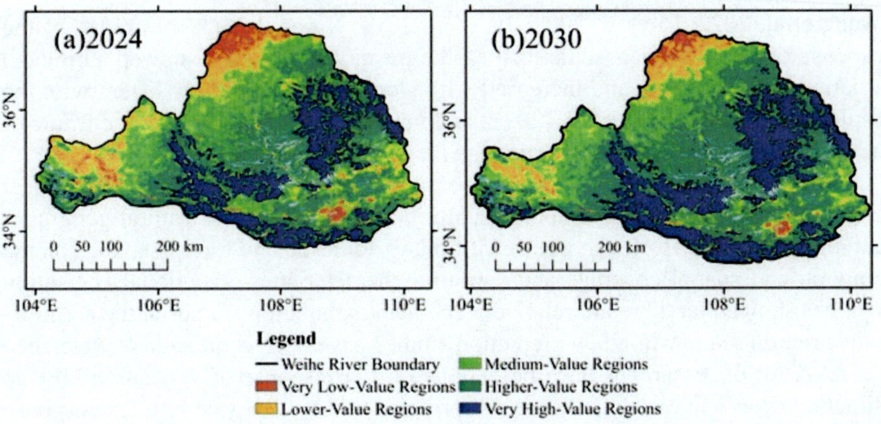

Fig. 18.9 Spatial distribution pattern prediction of NPP in the Weihe Watershed in 2024 and 2030

By 2024, the areas of very low-value regions, lower-value regions, and medium-value regions will have decreased by 6.30%, 9.48%, and 2.74%, respectively from 2018. Among them, the low level of NPP areas are mainly distributed in the northern part of the watershed and Xi'an and its surrounding areas. The area of the lower NPP areas has been significantly reduced, mainly in the northern and western parts of the basin, and the median NPP area expands northward as a whole. The areas with higher NPP and areas with high NPP are expanding northward, increasing by 9.56% and 8.96% respectively over 2018. By 2030, NPP will show a similar pattern of change.

In general, the overall vegetation growth conditions in the Weihe Watershed tend to be optimized. Among them, the areas of very low-value regions, lower-value regions, and medium-value regions will be reduced, and the overall spatial migration will be northward. The areas of higher-value regions and very high-value regions will be further increased and the spatial expansion will be northward and central.

18.4 Discussion and Conclusions

18.4.1 Discussion

In terms of the temporal trend of NPP, this study is consistent with the results of Wang et al. (2021) and Wang et al. (2020) who concluded that the NPP of vegetation in the Weihe River basin has been gradually increasing in recent years, and in terms of the spatial coverage of NPP, it is also consistent with the conclusions reached by the above-mentioned scholars. Judging from the response of NPP to climate factors, there is a positive correlation between NPP and temperature and precipitation in this study, accounting for as high as 70%. Some related studies also show that temperature and precipitation are generally positively correlated with NPP values in other regions

(Wang et al. 2022; Li et al. 2022; Jia et al. 2019; Xu et al. 2020). In terms of the response of NPP to topographic factors, the mean value of NPP showed a bimodal variation with elevation and increased with slope, which is not consistent with the results obtained by Wang et al. (2021) and may be influenced by the time frame of the study and the selection of parameters for different models.

The study of the characteristics, changes, and coverage of the natural geographical environment and its components is usually based on the law of natural geographical differentiation. However, a review of many domestic and foreign studies on the temporal and spatial coverage and variation characteristics of regional vegetation NPP has shown that there are relatively abundant simulation studies at the administrative region scale, while few prediction studies have been conducted at watershed scales. A lot of researches have been done on the response of vegetation NPP to climatic factors such as temperature and precipitation but few studies on its response to topographic factors. There have been many studies on vegetation NDVI prediction, but few studies on vegetation NPP prediction. Therefore, it is of great theoretical significance to simulate and predict the spatial and temporal coverage characteristics of watershed-scale vegetation NPP and explore its correlation with environmental factors.

Surface vegetation interacts with the atmosphere, hydrosphere, and soil sphere through material exchange and energy flow. The relationship between the degree and primary and secondary effects of many environmental factors on a vegetation system is very complex. This research focuses on analyzing the response of NPP to multiple single factors. In practice, NPP has relatively many influencing factors, such as the relationship between water and heat, the physiological and ecological properties of vegetation, and the interaction of climate factors in the models of climate change. At the same time, the temporal and spatial coverage and change characteristics of NPP in the Weihe Watershed are also closely related to human activities and economic construction, especially in the past 20 years, the Silk Road Economic Belt has been vigorously developed and the urbanization process has accelerated. The government has also issued a series of policies and regulations, such as Returning Farmland to Forest Program and the Hanjiang to Weihe River Project. These human activities will further affect the temporal and spatial coverage and change characteristics of NPP in the Weihe Watershed. It can be said that the change of NPP is a very complicated process. It is influenced not only by climate change and natural disasters, but also by the influence of land use policy, social, economic development and other human activities. Therefore, the follow-up will conduct in-depth research on the compound response of NPP to multiple factors, evaluate the contribution rate of various influencing factors to NPP, and better reveal the coupling mechanism of various factors to NPP.

18.4.2 Conclusions

Based on multi-source datasets, this paper experimented with the coupled application of CASA model and CA–Markov model, simulated and predicted the temporal and spatial coverage of NPP in the Weihe Watershed, and explored the response of NPP to environmental factors such as climate and topography based on related analysis methods. The main conclusions are as follows:

(1) Concerning temporality, the seasonality of NPP changes during the year is obvious; The trend of annual change of NPP is increasing, and it is predicted that NPP will keep increasing in the coming decade. There is a significant difference in spatial coverage of NPP, and the overall coverage characteristics of NPP are high in the south and low in the north, and high in the east and low in the west.
(2) The responses of NPP to temperature and precipitation in the Weihe Watershed are remarkable but varied. The response to temperature and precipitation is mainly positive correlation. More attention should be paid to heat preservation for the growth of vegetation in the northern part of the watershed, and more attention should be paid to the moisturizing effect for the growth of vegetation in the western part of the watershed.
(3) NPP in the Weihe Watershed responds significantly to topographical factors, but there are differences. NPP presents a bimodal trend with increasing altitude, and it keeps increasing with increasing slope. At the same time, the NPP of the north and west aspects are higher than that of the south and east aspects.
(4) The CASA model and the CA–Markov model have good coupling, and the prediction accuracy reaches 0.8776. It shows that the simulation based on CASA model and the prediction based on CA–Markov model can well evaluate the vegetation growth status and evolution law of the watershed.

References

Barnard JC, Powell DM (2002) A comparison between modeled and measured clear-sky radiative shortwave fluxes in arctic environments, with special emphasis on diffuse radiation. J Geophys Res 107(D19):9–10

Bian J, Li A, Deng W (2010) Estimation and analysis of net primary productivity of Ruoergai wetland in China for the recent 10 years based on remote sensing. Procedia Environ Sci 2:288–301

Bradford JB, Hicke JA, Lauenroth WK (2005) The relative importance of light-use efficiency modifications from environmental conditions and cultivation for estimation of large-scale net primary productivity. Remote Sens Environ 96(2):246–255

Cao M, Prince SD, Small J (2004) Remotely sensed interannual variations and trends in terrestrial net primary productivity 1981–2000. Ecosystems 7:233–244

Chen F, Shen Y, Li Q, Guo Y, Xu L (2011) Spatio-temporal variation analysis of ecological systems NPP in China in past 30 years. Geograph Sci 31:1409–1414

Defosse GE, Bertiller MB (1991) Comparison of four methods of grassland productivity assessment based on Festuca pallescens Phytomass Data. J Range Manag 44(3):199–203

Doraiswamy PC, Moulin S, Cook PW, Stern A (2003) Crop yield assessment from remote sensing. Photogramm Eng Remote Sens 69(6):665–674

Field CB, Randerson JT, Malmström CM (1995) Global net primary production: combining ecology and remote sensing. Remote Sens Environ 51(1):74–88

Field CB, Behrenfeld MJ, Randerson JT, Falkowski P (1998) Primary production of the biosphere: integrating terrestrial and oceanic components. Science 281(5374):237–240

Funk CC, Brown ME (2006) Inter-Seasonal NDVI change projections in semi-arid Africa. Remote Sens Environ 101:549–256

Guan X, Shen H, Gan W, Yang G, Wang L, Li X, Zhang LA (2017) A 33-Year NPP monitoring study in southwest china by the fusion of multi-source remote sensing and station data. Remote Sens 9(10):1082

Gonçalves RRV, Jr Zullo J, Romani LAS, Nascimento CR, Traina AJM (2012) Analysis of NDVI time series using cross-correlation and forecasting methods for monitoring Sugarcane fields in Brazil. Int J Remote Sens 33(15):4653–4672

Ji L, Peters AJ (2004) Forecasting vegetation greenness with satellite and climate data. IEEE Geosci Remote Sens Lett 1(1):3–6

Jia J, Liu H, Lin Z (2019) Multi-time scale changes of vegetation NPP in six provinces northwest China and its response to climate change. Acta Ecological Sinica 39(14):5058–5069

Jones JW, Hoogenboom G, Porter CH, Boote KJ, Batchelor WD, Hunt LA, Wilkens PW, Singh U, Gijsman AJ, Ritchie JT (2003) The DSSAT cropping system model. Eur J Agron 18(3–4):235–265

Kalubarme MH, Potdar MB, Manjunath KR, Mahey RK, Siddhu SS (2003) Growth profile based crop yield models: a case study of large area wheat yield modelling and its extendibility using atmospheric corrected NOAA AVHRR data. Int J Remote Sens 24(10):2037–2054

Kong F, Dong Q, Xiang K, Yin, Z, Li, Y, Liu J (2019) Spatiotemporal variability of remote sensing ocean net primary production and major forcing factors in the tropical eastern Indian and Western Pacific Ocean. Remote Sens 11(4):391

Lai C, Li J, Wang Z, Wu X, Zeng Z, Chen X, Lian Y, Yu H, Wang P, Bai X (2018) Drought-induced reduction in net primary productivity across mainland China from 1982 to 2015. Remote Sens 10(9):1433

Lewis JE, Rowland J, Nadeau A (1998) Estimating Maize production in Kenya using NDVI: some statistical considerations. J Remote Sens 19(13):2609–2617

Li Z, Wang X, Xu Y, Wen L, Huang L (2022) Changes of net primary productivity of Shangri-La Vegetation in Northwest Yunnan from 1996 to 2015. Chin J Acta Ecol Sinica 42(1):266–276

Li X, Liang H, Cheng W (2020) Spatio-temporal variation in AOD and correlation analysis with PAR and NPP in China from 2001 to 2017. Remote Sens 12(6):976

Luo Z, Wu W, Yu X, Song Q, Yang J, Wu J, Zhang H (2018) Variation of net primary production and its correlation with climate change and anthropogenic activities over the Tibetan Plateau. Remote Sens 10(9):1352

Naeem S, Zhang Y, Tian J, Qamer FM, Latif A, Paul PK (2020) Quantifying the impacts of anthropogenic activities and climate variations on vegetation productivity changes in China from 1985 to 2015. Remote Sens 12(7):1113

Nemani RR, Keeling CD, Hashimoto H, Jolly WM, Piper SC, Tucker CJ, Myneni RB, Running SW (2003) Climate-driven increases in global terrestrial net primary production from 1982 to 1999. Science 300(5625):1560–1563

Neumann M, Moreno A, Thurnher C, Mues V, Härkönen S, Mura M, Bouriaud O, Lang M, Cardellini G, Thivolle-Cazat A, Bronisz K, Merganic J, Alberdi I, Astrup R, Mohren F, Zhao M, Hasenauer H (2016) Creating a regional MODIS satellite-driven net primary production dataset for European forests. Remote Sens 8(7):554

Neumann M, Zhao M, Kindermann G, Hasenauer H (2015) Comparing MODIS net primary production estimates with terrestrial national forest inventory data in Austria. Remote Sens 7(4):3878–3906

Potter CS, Randerson JT, Field CB, Matson PA, Vitousek PM, Mooney HA, Klooster SA (1993) Terrestrial ecosystem production: a process model based on global satellite and surface data. Global Biogeochem Cycles 7(4):811–841

Prasad AK, Chai L, Singh RP, Kafatos M (2006) Crop yield estimation model for iowa using remote sensing and surface parameters. Int J Appl Earth Obs Geoinf 8(1):26–33

Ruimy A, Saugier B, Dedieu G (1994) Methodology for the estimation of terrestrial net primary production from remotely sensed data. J Geophys Res 99(D3):18515–18521

Shang E, Xu E, Zhang H, Liu F (2018) Analysis of spatiotemporal dynamics of the Chinese vegetation net primary productivity from the 1960s to the 2000s. Remote Sens 10(6):860

Wang L, Yu D, Liu Z, Yang Y, Zhang J, Han J, Mao Z (2018) Study on NDVI changes in Weihe Watershed based on CA-Markov model. Geol J 53:435–441

Wang Y, Xu X, Huang L, Yang G, Fan L, Wei P, Chen G (2019) An improved CASA model for estimating winter wheat yield from remote sensing images. Remote Sens 11(9):1088

Wang X, Wu M, Ma H, Guo Q (2020) Analysis of long-term series NPP estimation and spatiotemporal variation characteristics in Weihe River Basin. J Northw Forest Univ 35(6):205–211

Wang L, Zhang H, Liu Z, Zhang S, Kong J, Gao L (2021) A coupling model of net primary productivity pattern simulation and prediction. Geomat Inf Sci Wuhan Univ 46(11):1756–1765

Wang L, Ding H, Liu Z, Zhang S, Kong J (2022) Spatiotemporal change of NPP based on CASA model and its response to climate change in Jing river basin. Res Soil Water Conserv 29(1):190–196

Xu J, Chen H, Shang S, Yang H, Zhu G, Liu X (2020) Response of net primary productivity of Tibetan Plateau vegetation to climate change based on CEVSA model. Arid Land Geogr 43(3):592–601

Yu T, Sun R, Xiao Z, Zhang Q, Liu G, Cui T, Wang J (2018) Estimation of global vegetation productivity from global land surface satellite data. Remote Sens 10(2):327

Zhang R, Zhang X (2008) The evolution of the ecological environment in Shaanxi part of Weihe River Basin in recent 50 years. J Arid Land Resour Environ 22:37–42

Zhang M, Lin H, Sun H, Cai Y (2019) Estimation of vegetation productivity using a landsat 8 time series in a heavily urbanized area, central China. Remote Sens 11(2):133

Zheng Z, Zhu W, Zhang Y (2020) Direct and lagged effects of spring phenology on net primary productivity in the Alpine grasslands on the Tibetan Plateau. Remote Sens 12(7):1223

Zhu W, Chen Y, Pan Y, Li J (2004) Estimation of light utilization efficiency of vegetation in China based on GIS and RS. Geomat Inf Sci Wuhan Univ 29:694–698

Zhu W, Chen Y, Xu D, Li J (2005) Advances in terrestrial net primary estimation models. Chin J Ecol 24:296–300

Zhu W, Pan Y, Liu X, Wang A (2006) Spatio-temporal distribution of net primary productivity along the northeast China transect and its response to climatic change. J Forest Res 17(2):93–98

Zhu W, Pan Y, Zhang J (2007) Estimation of net primary productivity of Chinese terrestrial vegetation based on remote sensing. Chin J Plant Ecol 31(3):413–442

Chapter 19
Determination of Water-Soluble Cl^- and SO_4^{2-} in Soil by Chemical Analysis and Ion Chromatography

Liu Yang and Yin-Fu Wang

Abstract Titration and colorimetric chemical analysis methods are usually used to test the water-soluble Cl^- and SO_4^{2-} in soil. Although this method has been applied maturely, it also has some characteristics, such as complex operation, long analysis cycle, great influence of detailed process on analysis results, and requirement limit on sample content range. In this paper, the water-soluble Cl^- and SO_4^{2-} in standard soil samples were determined by chemical analysis and ion chromatography. The technical points of chemical analysis methods were put forward, and the possibility of using ion chromatography was discussed. The results show that the water-soluble Cl^- and SO_4^{2-} in soil are determined by ion chromatography. Under the selected test conditions, the mass concentrations of Cl^- and SO_4^{2-} are in the range of 1–100 mg/L, and the correlation coefficient R (Cl^-) = 0.9998, R (SO_4^{2-}) = 0.9997. The detection limits for chloride ions were 0.005 μg/ml and 0.016 μg/ml for sulfate. ASA14, ASA16, ASA17 and ASA18 were determined by the experimental method, and the accuracy and precision of the ion chromatography test were verified. The relative error of the test results was 0.7–8.3%, and the relative standard deviation was 0.8–2.8%. The determination of water-soluble Cl^- and SO_4^{2-} in soil by ion chromatography is feasible. The use of ion chromatography reagent is small, reduce the impact on the environment and human body, short test time, can better adapt to a large number of samples test analysis.

Keywords Water-soluble Cl^- and SO_4^{2-} in soil · Chemical analysis methods · Ion chromatography

Saline-alkali soil is widely distributed in China's cultivated land, which seriously restricts agricultural development and rural construction (Zhao et al. 2020). The content of water-soluble salt in soil is an important attribute of saline-alkali soil and an obstacle to crop growth. High salt content in soil will affect the soil quality in the tilled layer by affecting the formation of soil aggregates (Feng et al. 2020) and the

L. Yang (✉) · Y.-F. Wang
Shenyang Center, China Geological Survey, Shenyang 110034, Liaoning, China
e-mail: 570108151@qq.com

H. Xu (ed.), *Proceedings of the 5th International Symposium on Water Resource and Environmental Management*, Environmental Science and Engineering,
https://doi.org/10.1007/978-3-031-31289-2_19

migration of soil water (Liao et al. 2021; Wang et al. 2019), and at the same time affect the fertilizer application (Tao et al. 2020), crop growth and vegetation community distribution characteristics (Liu et al. 2021), thus restricting the improvement of agricultural productivity. These salts are mainly including the chlorides and sulfates of calcium, magnesium, sodium, potassium. Chloride ion (Cl^-) and sulfate ($SO_4{}^{2-}$) are the main components of soil soluble salt, and the essential index of anion content monitoring for soil (Jiang et al. 2016).

The technical methods of the determination of Cl^- and $SO_4{}^{2-}$ in water-soluble salt in soil was discussed in this paper.

Referring to the relevant standards for analysis of water-soluble salts in soil, Analysis of Water-soluble Salts in Forest Soil, LY/T 1251-1999, and detection of soil, NY/T 1121-2006, Cl^- could be determined by silver nitrate titration, $SO_4{}^{2-}$ could be determined by titration, colorimetric method, and mass method.

With the development of technology and analytical instrument, the determination of Cl^- and $SO_4{}^{2-}$ by ion chromatography has been widely used in water quality analysis, energy geology, environmental detection, food safety and other fields (Hu et al. 2018; Ma et al. 2019; Wang et al. 2012; Li et al. 2012; Yang et al. 2012). In practical application, ion chromatography has the advantages of fast, simple, sensitive, small sample amount required, good selectivity, and can analyze a variety of components at the same time. Especially for the determination of inorganic anions, it has other advantages that cannot be replaced.

Although the traditional chemical analysis method of water-soluble anion content in soil is mature, it has many complicated operations. If the determination of Cl^- and $SO_4{}^{2-}$ in water soluble salt can be carried out by ion chromatography, the speed of sample analysis will be greatly improved to meet the requirement of rapid analysis.

In this paper, the national standard material is taken as the research sample, the technical key points of traditional testing methods are summarized, and the feasibility of the determination of Cl^- and $SO_4{}^{2-}$ in soil water-soluble salt by ion chromatography is analyzed through the comparison of sample test results, to provide scientific basis and technical support for improving the testing efficiency of large quantities samples.

19.1 Experiment

19.1.1 Methods for Chemical Analysis (See LY/T 1251-1999 for Materials and Procedures)

(1) Determination of Cl^-

Silver nitrate titration method

Principle: Using $AgNO_3$ standard solution to titrate Cl^- with K_2CrO_4 as indicator, the reaction is as follows:

$Cl^- + Ag^+ \rightarrow AgCl\downarrow$ (White precipitate)

$CrO_4^{2-} + 2Ag^+ \rightarrow Ag_2CrO_4\downarrow$ (Brown–red precipitate)
Although both AgCl and Ag_2CrO_4 are precipitates at room temperature, the solubility of AgCl ($1.5*10^{-3}$ g/L) is smaller than that of Ag_2CrO_4 ($2.5*10^{-2}$ g/L). Therefore, when $AgNO_3$ is added to the solution, Cl^- first reacts with Ag^+ to form white AgCl precipitate. When all Cl^- in the solution is precipitated by Ag^+, Ag^+ reacts with K_2CrO_4 indicator to form brown–red Ag_2CrO_4 precipitate, and then the end point of titration is reached.

(2) Determination of SO_4^{2-}
Sulfate analysis is a complicated project in water-soluble salt analysis. The classical method is barium sulfate precipitation weighing method, which has the advantage of high accuracy. The difficulty lies in the large amount of soil leaching solution required (more than 100 ml), and it is only suitable for samples with relatively high SO_4^{2-} content (100 ml containing sulfate should be more than 20 mg, preferably 100 mg, if the concentration is below 10 mg, more leaching solution should be taken and the leaching solution should be concentrated and evaporated). The whole process of precipitation evaporation takes a long time, so the method mentioned above is now less used. EDTA indirect titration and turbidimetric method are widely used now.

①EDTA indirect titration method (optimum determination range is 50–300 μg/ml)

Principle: Excessive barium chloride solution is used to precipitate the SO_4^{2-} in the solution completely. To prevent the precipitation of $BaCO_3$, the solution to be tested must be acidified before adding $BaCl_2$ solution and heated to boiling to drive out CO_2. $BaCl_2$ solution was added while hot to promote the precipitation of $BaSO_4$ to form larger particles. Excessive Ba^{2+}, together with the existing Ca^{2+} and Mg^{2+} in the solution, were determined by EDTA standard solution with chromium black T indicator at pH = 10. It is worth mentioning that for making the end point obvious, a certain amount of Mg^{2+} should be added. So, the consumption of EDTA consists of three parts: Mg^{2+} added to make the titration endpoint obvious, Ba^{2+} in the solution to be tested, and Ca^{2+} and Mg^{2+} existing initially in the solution which can react with EDTA. The amount of EDTA consumed by Mg^{2+} to make the end point of titration obvious can be obtained by using blank calibration. The concentrations of Ca^{2+} and Mg^{2+} existing initially in the sample were measured in advance, and the amount of EDTA consumed by corresponding Ca^{2+} and Mg^{2+} in the solution to be tested could be obtained by calculation. Finally, the amount of Ba^{2+} in the solution to be tested can be calculated to find the SO_4^{2-} amount.

②Barium sulfate turbidimetric method (the best determination range is <40 μg/ml);

Principle: The sulfate in the solution to be tested and the added barium chloride ($BaCl_2$) crystals formed dispersed and stable barium sulfate ($BaSO_4$) suspension, and its absorbance (turbidity) was measured by colorimeter or turbidimeter.

According to the measurement results, the concentration of sulfate was found on the standard curve.

19.1.2 Ion Chromatography

Instruments, reagents and materials.

AQUION RFIC ion Chromatograph: ThermoFisher.

SPECORD 210 UV–Visible Spectrophotometer: Agilent.

HY-2 speed regulating multi-purpose oscillator: Ronghua instrument.

Centrifuge: Feige.

Cl^-, SO_4^{2-} standard solution (1000 mg/L) were purchased from Beijing North Weiye Metrology Technology Research Institute.

The experimental standard samples were purchased from Environmental Quality Supervision, Inspection and Testing Center of Ministry of Agriculture and Rural Affairs (Tianjin).

All reference substances are provided with a standard sample certificate.

Process parameter for chromatography.

Cl^-, SO_4^{2-} mixed standard curve. The 1000 μg/ml mixed standard solution (CUST-96252) were purchased from Beijing North Weiye Institute of Metrology Technology (Table 19.1).

Accurately remove 0.00 ml, 0.10 ml, 0.50 ml, 1.00 ml, 5.00 ml, 10.00 ml standard solution into a set of 100 ml volumetric bottles, dilute with water, volume to 100 ml/bottle and mix well. Formulated into 6 different concentrations of mixed standard solution.

Table 19.1 Working conditions of IC

Process parameter	Value
Eluent buffer	30%mmolKOH
IC electric current	75 mA
Eluent flow rate	1.0 ml/min
Partial loop fill	25 μl
Target of measurement	peak area
Time	12 min
Chromatographic column	AS11-HC
Guard column	AG11-HC
Suppressor	AERS500- 4 mm
Detector detection parameter	Conductivity

Quality control. Quality control was carried out by measuring the standards. The mixed standard solutions GNM-M031909-2013 and GNM-M031912-2013 were purchased from the National Analysis and Testing Center for Non-ferrous Metals and Electronic Materials (the reference values of Cl^- and SO_4^{2-} were 10 μg/ml and 100 μg/ml). Standards are used to monitor the stability of instruments during analytical testing to ensure the accuracy of measurement data.

19.1.3 *Sample Treatment*

Weigh the 50.0 g (10 mesh sieves filtered) air-dried soil sample into a 300 ml plastic extraction bottle, add 250 ml CO_2-free pure water that has been boiled and cooled to room temperature, cover the bottle, and put it into an oscillator, oscillate at 220 r/min for 3 min. The extract was filtered by 0.45 μm filter membrane to obtain the clarified liquid to be tested.

19.2 Results and discussion

19.2.1 *Determination of Water-Soluble Cl^- and SO_4^{2-} in Soil by Chemical Method*

Results discussion and summary of precautions in operation (Table 19.2).

According to the experimental method, the four-soil available state standards were measured, and the parallel determination was made for five times. The results were

Table 19.2 The results of chemical determination of water-soluble Cl^- and SO_4^2 in available state reference materials in soil

Sample No	Property	Cl^- μg/g	Silver nitrate titration		SO_4^{2-} μg/g	EDTA titration		Turbidimetry	
			Mean value	RSD% ($n = 5$)		Mean value	RSD	Value	RSD% ($n = 5$)
ASA14	Heilongjiang black soil	0.030	0.028	1.9	0.118	0.116	3.1	0.122	4.5
ASA16	Henan tidal soil	0.036	0.034	3.8	0.144	0.140	2.1	0.139	3.9
ASA17	Xinjiang chestnut soil	0.048	0.041	2.4	0.151	0.155	2.4	0.159	3.8
ASA18	Jilin brown soil	0.027	0.03	3.1	0.112	0.115	1.9	0.117	5.2

all within the scope given by the standard certificate. The test results are good. What should be noted in the operation process is detailed below:

(1) Titrations of Cl^- with AgNO should be carried out in neutral solutions because the following reactions occur in acidic environments: $CrO_4^{2-} + H^+ \rightarrow HCrO_4^-$ which reducing the sensitivity of K_2CrO_4 as an indicator. If in alkaline environments: Ag+ +OH $\rightarrow$ AgOH$\downarrow$, The concentration of Ag^+ in AgOH saturated solution is lower than that in Ag_2CrO_4, so AgOH will precipitate out before Ag_2CrO_4. This will result in no brown–red precipitate at the end of the titration of Cl^-, which will affect the determination of Cl^-. Therefore, we recommend that the solution after the determination of CO_3^{2-} and HCO_3^- be used for the determination of Cl^-. The endpoint will be easier to identify if the titration is done in yellow light. If the color of the leaching solution is dark and it is difficult to distinguish the color change of the end point, potentiometric titration can be used instead.
(2) The EDTA indirect titration method for the determination of SO_4^{2-} is relatively simple. However, the addition amount of barium magnesium mixture is not easy to determine, and the end point of titration is difficult to master. If the dosage of barium magnesium mixture is insufficient, it will lead to wrong measurement results. Therefore, it is necessary to predict the approximate content of sulfate in the sample properly before analysis to determine the dosage of barium magnesium mixture, to obtain more accurate determination results.
(3) Barium sulfate turbidimetric method for the determination of SO_4^{2-} is quick and convenient. The disadvantage is the result is easily affected by precipitation conditions, such as uniform suspension without forming conditions. This will affect the accuracy of the results.

19.2.2 Determination of Cl^- and SO_4^{2-} by ion Chromatography

Standard curve and detection limit. Determination of standard solution series under stable instrumental conditions. With the mass concentration of chloride ion and sulfate as the abscissa and the corresponding peak area as the ordinate, the standard curve was drawn, as shown in Fig. 19.1. The results show that the linear relationship between the standard curves of Cl^- and SO_4^{2-} is good when the mass concentration of Cl^- and SO_4^{2-} is within 1–100 μg/ml. Correlation coefficient $R^2_{(Cl-)} = 0.9998$ $R^2_{(SO42-)} = 0.9997$. Under the same conditions, the blank solution was tested continuously for 7 times according to the experimental method. The detection limit of the detection method is calculated (three times the standard deviation). The detection limit of chloride ion is 0.005 μg/ml, the detection limit of sulfate is 0.016 μg/ml. Four times the detection limit was taken as the lower limit of determination, The lower limit for the determination of chloride ions was 0.02 μg/ml, and the detection limit for sulfate was 0.064 μg/ml (Fig. 19.2).

Precision and accuracy. According to the experimental method, the aqueous solution of the available soil standards ASA14, ASA16, ASA17 and ASA18 was extracted. The water-soluble Cl^- and SO_4^{2-} were determined under stable instrumental conditions, The results are shown in Figs. 19.3–19.6 and Table 19.3. The four-soil available state standards were measured six times parallelly according to the experimental method. All the results are within the scope given by the standard certificate. The test results are good. The relative deviation of the measurement results

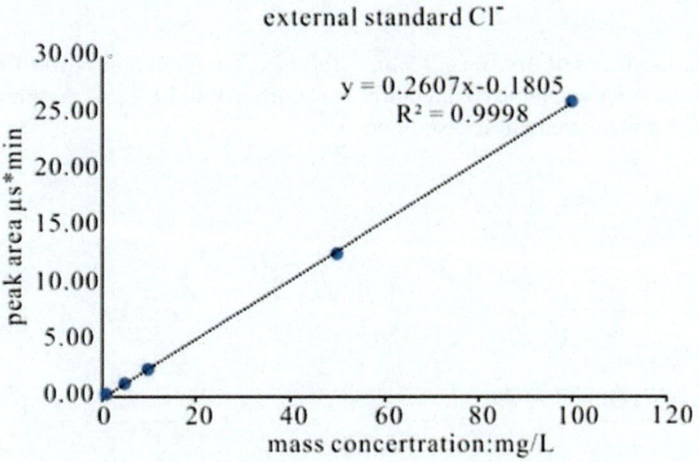

Fig. 19.1 Calibration curve of Cl^-

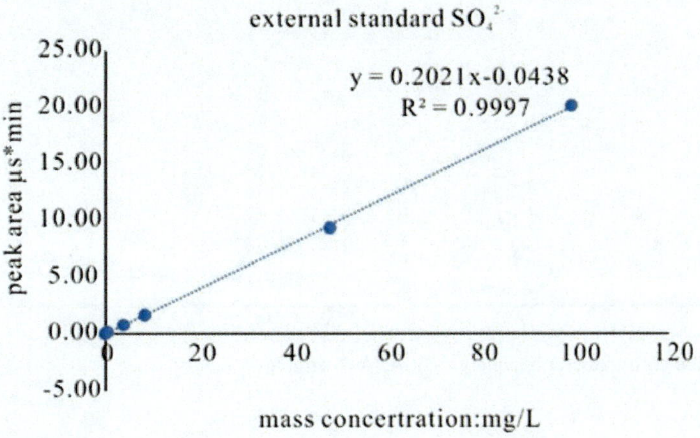

Fig. 19.2 Calibration curve of SO_4^{2-}

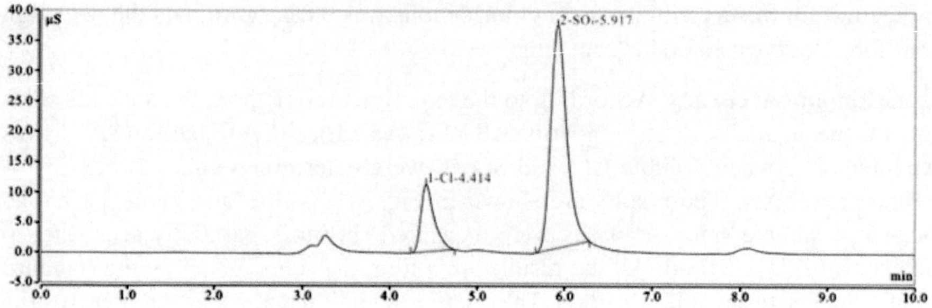

Fig. 19.3 Ion chromatographic analysis of ASA14 sample

met the requirements of *Analysis of Water-Soluble Salinity in Forest Soil LY/T1251-1999*, and the relative standard deviation was within 1.1–2.8%, so the precision of the instrument measurement is also good.

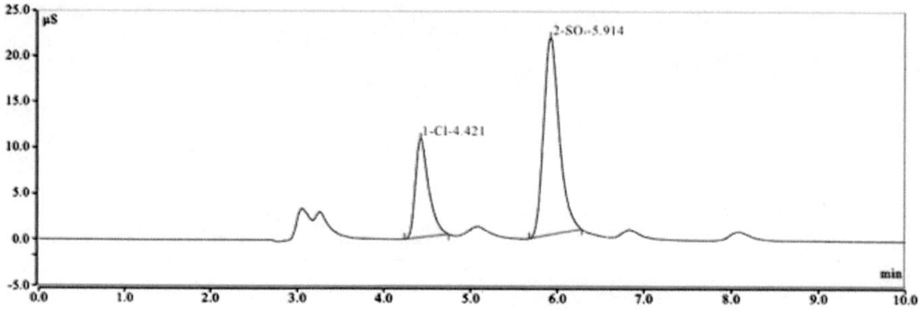

Fig. 19.4 Ion chromatographic analysis of ASA16 sample

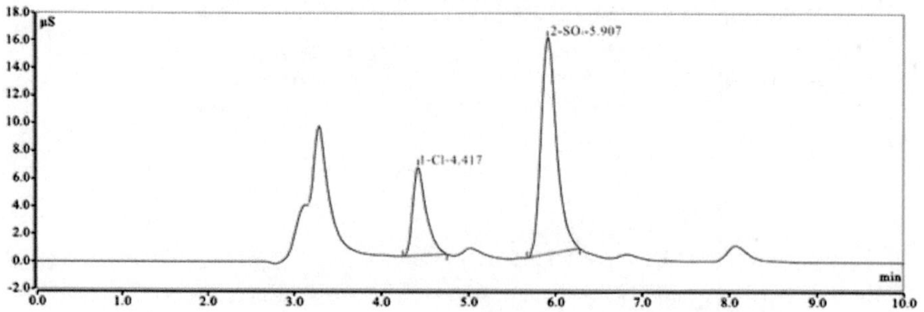

Fig. 19.5 Ion chromatographic analysis of ASA17 sample

Table 19.3 Determination of Water-soluble Cl^- and SO_4^{2-} in available state reference materials in soil by ion chromatography

Sample No	Cl^- content μg/g	Ion chromatography			SO_4^{2-} content μg/g	Ion chromatography		
		Mean value	RSD% ($n = 6$)	RE%		Mean value	RSD% ($n = 6$)	RE%
ASA14	0.030	0.029	1.1	3.3	0.118	0.117	1.1	0.9
ASA17	0.048	0.044	0.8	8.3	0.151	0.150	1.4	0.7
ASA18	0.027	0.029	1.1	7.4	0.112	0.113	1.9	0.9
ASA16	0.036	0.035	2.8	2.8	0.144	0.141	1.1	2.1

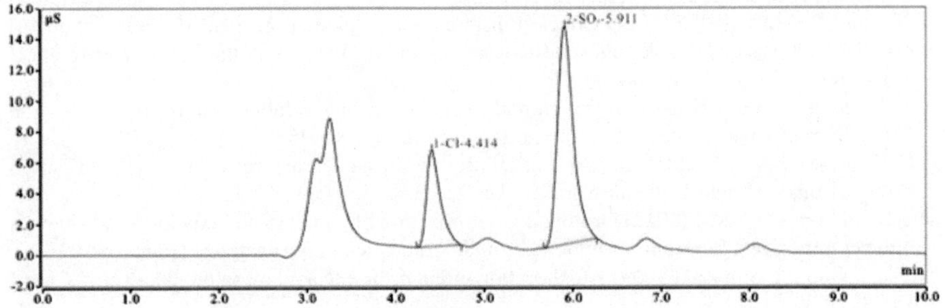

Fig. 19.6 Ion chromatographic analysis of ASA18 sample

19.3 Conclusion

The traditional chemical analysis method for the determination of soil water-soluble Cl^- and SO_4^{2-} has corresponding standards for reference. The technology and methods are mature and reliable, However, due to the relatively complex operation process, and it is necessary to select appropriate detection methods according to different sample contents for the determination of the corresponding sulfate. Therefore, the traditional analysis method has higher technical requirements for personnel. Compared with traditional chemical method, ion chromatography is quick and simple, which greatly improves the analysis efficiency and saves the analysis cost. It is suitable for mass sample analysis. In the determination of water-soluble Cl^- and SO_4^{2-} by ion chromatography, the technical indexes such as precision, accuracy and detection limit all meet the quality requirements of analysis in the standard for geochemical evaluation of soil quality. We hope to apply for the relevant standard method to provide technical support for soil environmental protection and pollution control through laboratory collaboration analysis, and hope that this method can play an important role in the third national soil census in China.

References

Feng YQ, He TH, Chen XQ (2020) Study on the relationship between plant diversity and soil texture and salinity of saline meadow community. ACTA AGRESTIA SINICA 28(6):1682–1689

Hu JZ, Wang L, Liu J (2018) Determination of water soluble sulfate in soil by inductively coupled plasma atomic emission spectrometry. Metall Anal 38(11):12–17

Jiang Q, Han Y, Sun XL (2016) Study on the determination and its difference analysis of chloride and sulfate in different soils by ion chromatography and capillary electrophoresis. Soils 48(2):343–348

Li QX, Liu YX, Chen WM (2012) The determination of seven anions in water-extractable soil samples by ion chromatography. Geophys Geochem Explor 3:418–421

Liao H, Li XW, Chen JY (2021) Effects of different salt contents of undisturbed saline soil on soil moisture characteristic curves. Water Sav Irrig 1:7–13

Liu SL, Ding XD, Zheng DF (2021) Effect of different plants plantation on amelioration of uncultivated saline wasteland, soils phosphorus fraction and availability in the yellow river delta. J Soil Water Conserv 35(1):278–284

Ma BB, Jiang Y, Luo Y (2019) Determination of ten water-soluble anions in soil by ion chromatography with ultrasonic extraction. Soils 51(6):1253–1256

Tao JY, Yang JS, Yao RJ (2020) Effects of soil salinity on nitrogen transformation in Hetao irrigation district of inner Mongolia. China Soils 52(4):802–810

Wang HB, Xie J, Ding XM (2012) Orthogonal optimization of soil extraction method and analysis of the common inorganic anions by ion chromatography. Chin J Anal Lab 31(6):103–107

Wang QZ, Wang Y, Sun ZM (2019) Amelioration effect of humic acid on saline-alkali soil. Chin J Appl Ecol 30(4):1227–1234

Yang CX, Li CH, Zhao YB (2012) IC determination of inorganic anions and cations in soil. PTCA(Part B:Chem Anal) **48**(10): 1199–1202

Zhao YY, Zhang ZY, Tang XD (2020) Analysis of high content water-soluble salt cation in saline-alkali soil by X-ray fluorescence spectrometry. Spectrosc Spectr Anal 40(5):1467–1472

Chapter 20
Hydrological Change and Probable Ecosystem Impacts Under a Climate Change in the Kunhar River, Pakistan

Shan-E-Hyder Soomro, Xiaotao Shi, Jiali Guo, Yinghai Li, Caihong Hu, Zhu Chunyun, Jiahui Gu, Ao Li, Shenghong Liu, Yu Guo, and Nida Rabab

Abstract Nowadays Pakistan is suffering from water stress, and its water supplies are under threat, because of the unpredictability of climate change. Hydrologiska Byrns Vattenbalansavdelning (HBV) model was used to calibrate and validate the anticipated flow fluctuation in Kunhar River in the face of persistent climatic variations. Overall Nash–Sutcliffe yield of 0.87 and 0.79 was found to be acceptable for the HBV work ability through the phases of overall calibration and validation, respectively. The research shows that future changes are predicted in the study region's streamflow intensity, duration, frequency, timing, and rate. For the period 2011–2099, all scenarios predict a hydrologic change relative to historical values for the majority of flow characteristics (1975–2020). Varying hydrologic and temperatures could have an impact on the fish population in the area being analyzed. The findings of this study have the potential to improve water resources and aquatic environment management by supplying useful and actionable information.

Keywords Climate change · Ecological management · HBV modeling · Streamflow · Kunhar river · Pakistan

S.-E.-H. Soomro · X. Shi (✉) · J. Guo (✉) · Y. Li · Z. Chunyun · J. Gu · A. Li · S. Liu · Y. Guo
College of Hydraulic and Environmental Engineering, China Three Gorges University, Yichang 443002, China
e-mail: fishlab@163.com

J. Guo
e-mail: jiali.guo@ctgu.edu.cn

S.-E.-H. Soomro · C. Hu
College of Water Conservancy Science and Engineering, Zhengzhou University, Zhengzhou 450001, China

N. Rabab
Institute of Business Management Korangi Creek, Karachi Sindh 75190, Pakistan

H. Xu (ed.), *Proceedings of the 5th International Symposium on Water Resource and Environmental Management*, Environmental Science and Engineering,
https://doi.org/10.1007/978-3-031-31289-2_20

20.1 Introduction

Hydrological systems are highly significant because of their effect on economic and ecological development. The scarcity of fresh water in Pakistan is a severe problem since the average annual supply has declined from 5,000 m^3 per person in 1952 to 1,100 m^3 in 2014 (Arif et al. 2022). A UN report defines nations with water shortage as those having annual per capita availability of less than 1,000 cubic meters of water (Dhakal et al. 2022). The dynamics of a basin's hydrological cycle are often affected by factors such as location, climate, and human interference. Variations in climatic factors such as temperature, precipitation, humidity, and wind have been observed by numerous researchers (Wang et al. 2022, Zan et al. 2022). Recently, (Yifru et al. 2022) have begun to use GCMs to explore the impacts of forthcoming climatic alteration on freshwater properties worldwide. There is a high correlation between river biodiversity and ecosystem sustainability, and the natural flow regime of river systems (Widén et al. 2022) and in addition to controlling a wide variety of ecological and physiological phenomena (Litchman and Thomas 2023). Physical properties of river habitats, such as flow rate and water depth, are substantially influenced by processes such as nutrient cycling and sediment transport (Yuan et al. 2023). Hence, the stream management of a river is critical to the survival of aquatic, riparian, and wetland ecosystems along with the sustainability of their surrounding environment (Abebe et al. 2022). Changing flow regimes can cause stream ecosystems to lose native species, fail to recruit fish and micro-invertebrates, and introduce invasive alien species (Feio et al. 2023). For environment and biodiversity of the region, fish is the crucial part which played significant role (Prakash 2021). There are near 193 recognized fish species in the rivers of Pakistan, but only approximately 30 are considered economically significant inhabitants. These 30 species include: cyprinus carpio, Labeo rohita, Gibelion catla, Cirrhinus mrigala, Cirrhinus reba, Channa straita, Channa marulius, Sperata sarwari, Wallago attu, Rita rita, Bagarius baga (Usman et al. 2016). In relation to soil conservation and evapotranspiration fluxes, vegetation is an essential component of the watershed water cycle resulting in an essential part of the region's biodiversity. Hydrological alteration causes significant ecological hazards to Kunhar River aquatic and terrestrial species. Low fish biodiversity has been linked to Kunhar River temperature fluctuations and turbidity challenges (Soomro et al. 2022, 2023), and there's evidence that physical causes linked to climate change are responsible for the Kunhar River's water quality deteriorating. Based on ecology and social potential consequences of the changing climatic scenarios in the Kunhar River basin, prior analysis has emphasized upon hydrodynamic effects caused by fluctuations and catastrophic river discharge amounts. Regarding specific reasons, we need an in-depth investigation of consequences that climate change has on the river system ecology within that vicinity. Thus, this research is to determine and evaluate the relevant primary water related streamflow modification which may have significant consequences on aquatic organisms in the Kunhar River watershed under rapidly changing climatic conditions.

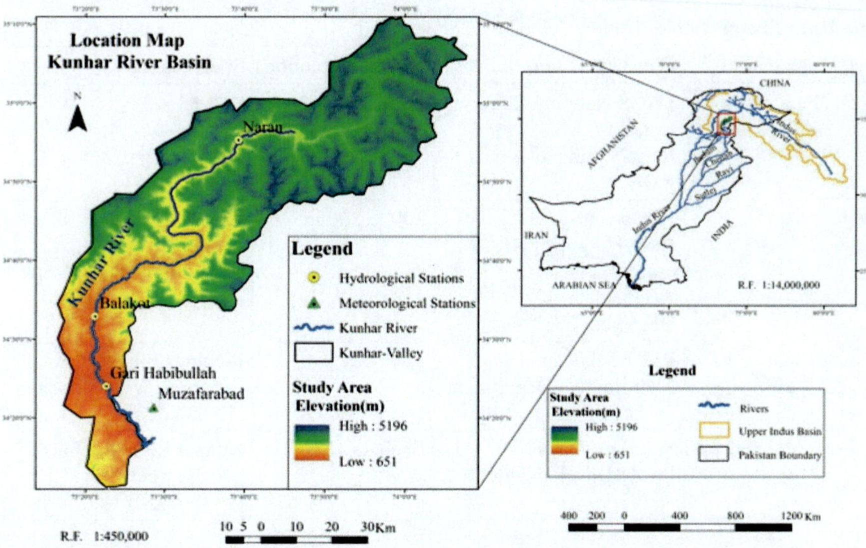

Fig. 20.1 Map representing kunhar river basin location used in the current experiment

20.2 Case of Kunhar River and Acquisition of Data

The research was assessed for the Kunhar River which stretches across Northern Pakistan from 34.2–35.1°N to 73.3–74.1°E. The overall area of Kunhar River is about (171 km) long, which exists in alpine valley located in Mansehra district. The approximately 2600 km^2 of drainage area and an elevation from 672 to 5192 m above sea level shown in Fig. 20.1. Also, the river water is crucial for domestic purposes which turn to contributes in region's economic growth. The hydrometeorological data were obtained from different sources mentioned in (Table 20.1).

The hydrosphere in a region can be adversely affected by land cover. Vegetation diversity and its density have significant effects on these processes (Zhao et al. 2022) shown in Fig. 20.2.

20.3 Materials and Methods

20.3.1 HBV Hydrological Model

The HBV model is a conceptual hydrology model that simulates the runoff mechanism in a watershed based on meteorological data like air temperature and potential evapotranspiration (Devia et al. 2015). The input data to construct the HBV model

Table 20.1 List of sources from which data were obtained

Type of the data	Data acquisition sources	Spatial resolution	Specification
Terrain data	USGS National Elevation Dataset (http://earthexplorer.usgs.gov/)	30 × 30 m	DEM
Landuse data	European Space Agency (ESA) http://ionia1.esrin.esa.int/ Access date: 12 December 2022	300 × 300 m	Land classification like water body, forest, agricultural, snow, crops, etc
Climate data	PMD-Pakistan Metrological Department	Each day	Precipitation, heat, solar radiation, Wind velocity (1975–2020)
Hydrological metrics	WAPDA-Water and Power Development Authority	Each day	Kunhar River discharge at daily time step (1975–2020)

comprised mainly quantifiable parameters, such as rainfall, temperature, and potential evaporation, so its uncertainty is minimal in most conditions (Ouatiki et al. 2020). The HBV hypothesis is feasible due to its simplicity and robust framework. The overall methodology is presented in Fig. 20.3.

20.4 Results and Discussion

20.4.1 Calibration and Validation Over Kunhar River

HBV model performance with the calibrated a predetermined set of variables or parameters adequately represents the hydrological features of the Kunhar River Basin for the period of 2000 to 2009 and simoultaneouly validated for the dataset from 2010 to 2016 for two different stream gauge stations named Gari-Habibullah and Naran in the Kunhar river basin. Repeated HBV model run calibration and validation for the two stream gauge stations for the Gari-Habibullah and Naran basins yielded efficiency levels 0.87 and 0.79 of Nash–Sutcliffe (NSE), respectively. The calibration and validation procedures brought to a remarkable enhancement in the correlation between the predicted and detected river flows. The model calibration accurately depicted the abundant flow in Gari-Habibullah in 2006 and the extreme discharge in 2007; whereas, during validation, the discharge was consistently unpredictable and wildly misleading in both 2011 and 2015. The estimated annual runoff from the simulation and evaluation for the validation periods of 2010–2016 was 488 and 495 mm, respectively (Table 20.2; Fig. 20.4).

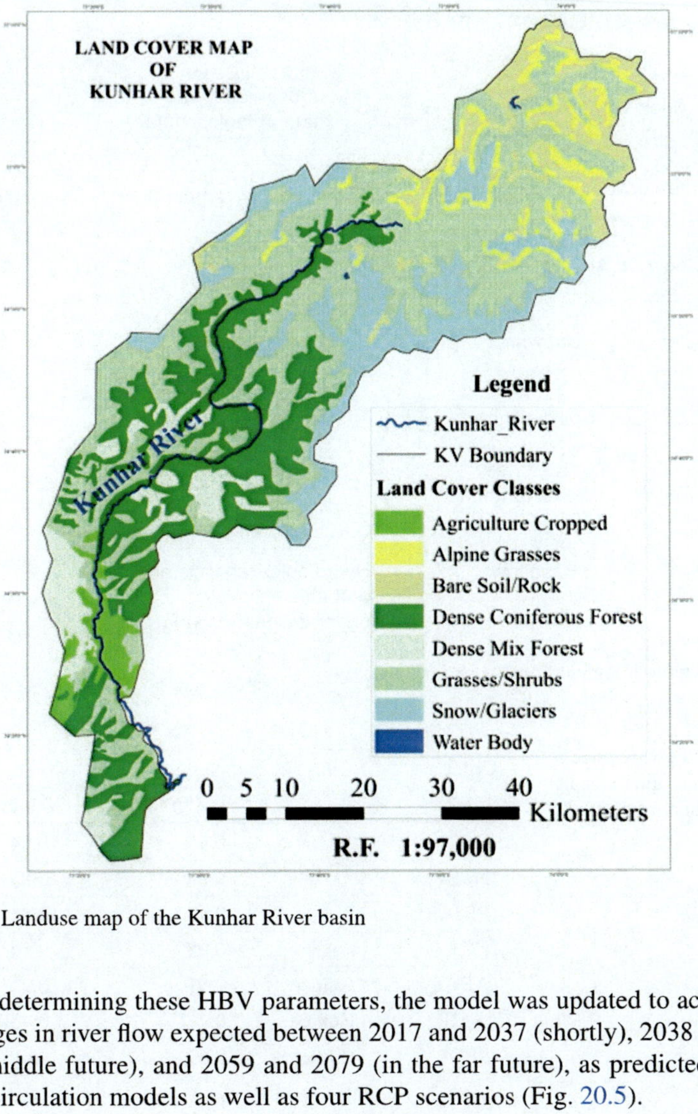

Fig. 20.2 Landuse map of the Kunhar River basin

After determining these HBV parameters, the model was updated to account for the changes in river flow expected between 2017 and 2037 (shortly), 2038 and 2058 (in the middle future), and 2059 and 2079 (in the far future), as predicted by four general circulation models as well as four RCP scenarios (Fig. 20.5).

20.4.2 *Measure of Hydrological Alteration and Ecological Implications*

Projected Changes in Streamflow. RCP 8.5 emission scenario maximum flow variability was computed (Fig. 20.6). For the long-term RCP 2.6 emission scenario, BCC-CSM2-MR is forecast to increase mid-term monsoon flow by 42%, while MPI-ESM1.2-HR is expected to decrease it by 19% in winter.

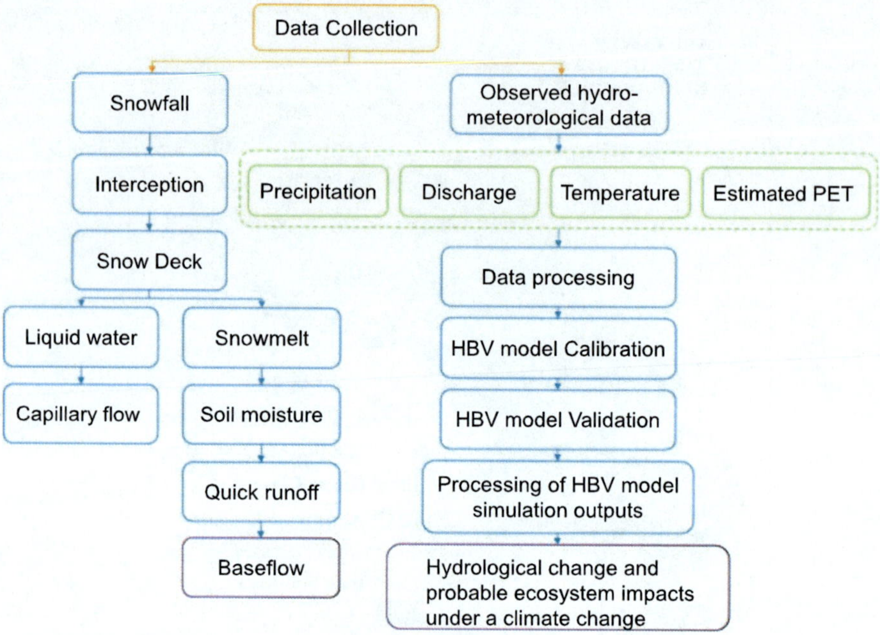

Fig. 20.3 Methodological framework for the study

Table 20.2 Performance indicators and outcomes in HBV for Kunhar River

Criteria	Units	Annually	
		Calibration	Validation
(R^2)		0.95	0.94
(NSE)		0.88	0.85
(PBIAS)	%	0.47	14.61
(CC)		0.95	0.94
(AE)	Cumec	0.01	0.24
(AAE)	Cumec	0.4	0.45
(SE)	Cumec	0.69	0.68

Note R^2 is coefficient of determination, NSE is Nash–Sutcliffe efficiency, PBIAS is a Percentage bias, CC is the correlation coefficient, AE is the average error, AAE is a average absolute error, and SE is a standard error

The RCP 4.5 emission scenario under MPI-ESM1.2-h is predicted a total 63% risen during next few months in monsoon season and 34% increase in extreme winter months. It was found that peak cold months will reduce by 21% under the RCP 6.0 emission scenario, while the CAMS-CSM1-0 model predicted a 76% increase in rainy season stream flow over multiple months. However, 149% predicted rise during monsoon under CAMS-CSM1-0 during initiation times in the distant future,

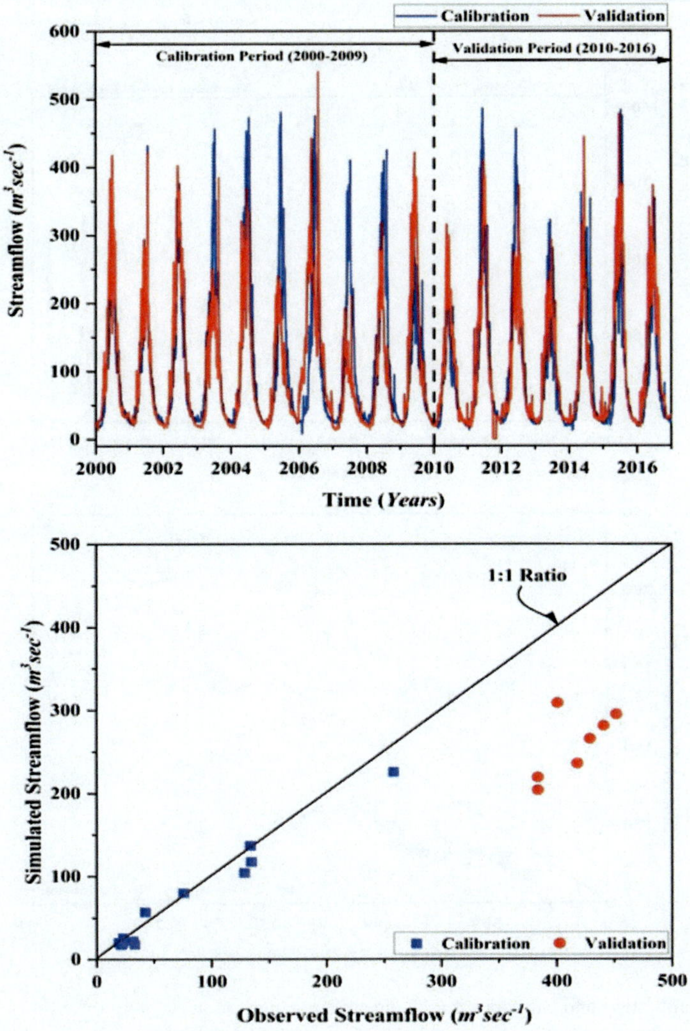

Fig. 20.4 Calibration and validation for Gari-Habibullah stream gauge

and MPI-ESM1.2-h predicted a decline of fall flows to 36% under the RCP 8.5 emission scenario. Analyses of water flow indicate a massive increase in the amount of water flowing during the warm months and a precipitous fall during the winter.

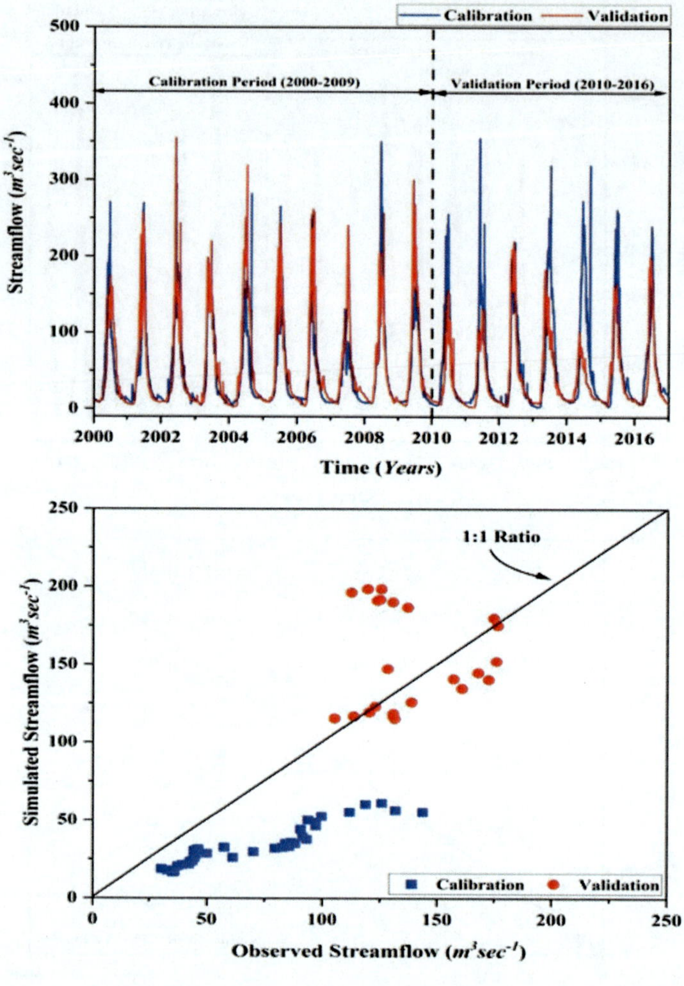

Fig. 20.5 Calibration and validation for Naran stream gauge

20.5 Conclusion

The HBV model was used to calibrate and validate the expected flow fluctuation in the Kunhar River in light of current climate change conditions. For the period 2011–2099, all scenarios predict a hydrologic change relative to historical values for the majority of flow metrics (1975–2020). It was predicted that in the mid-twentieth century, natural streamflow conditions will change. Based on the findings, it seems that a Nash–Sutcliffe efficiency of 0.87 is optimal for the calibration stage, while 0.79 is suitable for the validation stage. In addition, peak cold months will reduce by 21% under the RCP 6.0 emission scenario, while the CAMS-CSM1-0 model predicted a 76% increase in rainy season stream flow over multiple months. However, 149%

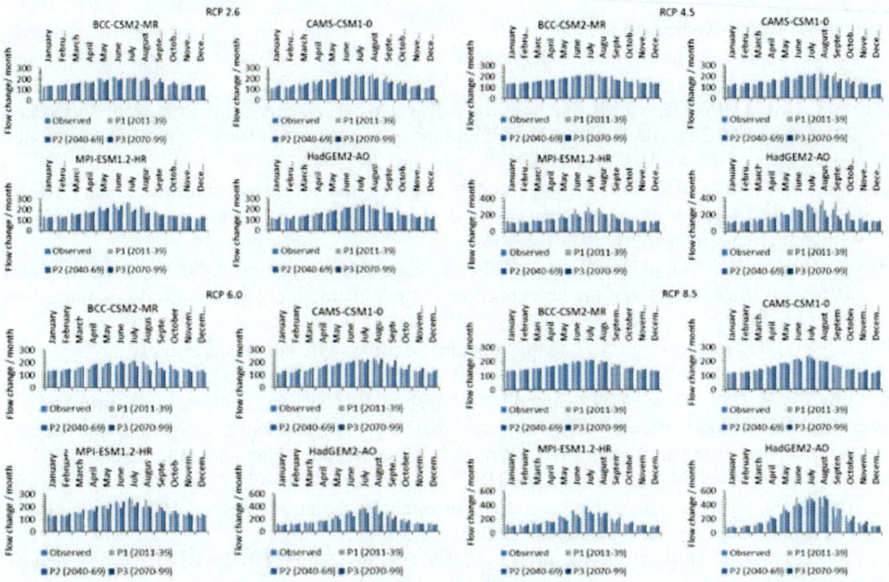

Fig. 20.6 Projected changes in streamflow

predicted rise during monsoon under CAMS-CSM1-0 during initiation times in the distant future, and MPI-ESM1.2-h predicted a decline of fall flows to 36% under the RCP 8.5 emission scenario. These changes may affect the Kunhar River's fish biodiversity, particularly during their reproductive and growth stages. This work helps stakeholders in policy and decision-making assess aquatic ecosystem vulnerability to hydrologic changes and manage water resources and aquatic ecosystems more effectively and sustainably.

Acknowledgements This work was supported by the National Natural Science Foundation of China (51922065, 52279069), Projects of the National Natural Science Foundation of China (52179018).

References

Abebe WB, Tilahun SA, Moges MM, Wondie A, Dersseh MG, Mcclain ME (2022) Environmental flow assessment and implications on sustainability of aquatic ecosystems in Ethiopia: a literature review on global and national evidences. Environ Dev 100758

Arif SMWK, Bannian B, Khan Q (2022) Water management and reservoirs in pakistan: an analysis of legal framework and contemporary challenges. Adv State Stud 1:19–24

Devia GK, Ganasri BP, Dwarakish GS (2015) A review on hydrological models. Aquat Procedia 4:1001–1007

Dhakal N, Salinas-Rodriguez SG, Hamdani J, Abushaban A, Sawalha H, Schippers JC, Kennedy MD (2022) Is desalination a solution to freshwater scarcity in developing countries? Membranes 12:381

Feio MJ, Hughes RM, Serra SR, Nichols SJ, Kefford BJ, Lintermans M, Robinson W, Odume ON, Callisto M, Macedo DR (2023) Fish and macroinvertebrate assemblages reveal extensive degradation of the world's rivers. Glob Change Biol 29:355–374

Litchman E, Thomas MK (2023) Are we underestimating the ecological and evolutionary effects of warming? Interactions with other environmental drivers may increase species vulnerability to high temperatures. Oikos 2023:E09155

Ouatiki H, Boudhar A, Ouhinou A, Beljadid A, Leblanc M, Chehbouni A (2020) Sensitivity and interdependency analysis of the hbv conceptual model parameters in a semi-arid mountainous watershed. Water 12:2440

Prakash S (2021) Impact of climate change on aquatic ecosystem and its biodiversity: an overview. Int J Biol Innov 3

Soomro S-E-H, Hu C, Boota MW, Ahmed Z, Chengshuai L, Zhenyue H, Xiang L, Soomro MHAA (2022) River flood susceptibility and basin maturity analyzed using a coupled approach of geo-morphometric parameters and swat model. Water Resour Manag 36:2131–2160

Soomro S-E-H, Shi X, Guo J, Hu C, Zwain HM, Liu C, Khan MZ, Niu C, Zhao C, Ahmed Z (2023) Appraisal of climate change and source of heavy metals, sediments in water of the Kunhar River watershed, Pakistan. Nat Hazards 1–19

Usman K, Pervaiz K, Hassan Z, Rehman HU, Ullah S, Haider I, Sadia H, Kiani BH, Khan MY, Nawaz A (2016) Exploring the fish fauna in water bodies of Hazara division, Khyber Pakhtunkhwa. Pakistan. J Entomol Zool Stud 4:848–851

Wang D, Feng H, Zhang B, Wei Z, Tian Y (2022) Quantifying the impacts of climate change and vegetation change on decreased runoff In China's yellow river basin. Ecohydrol Hydrobiol 22:310–322

Widén Å, Malm Renöfält B, Degerman E, Wisaeus D, Jansson R (2022) Environmental flow scenarios for a regulated river system: projecting catchment-wide ecosystem benefits and consequences for hydroelectric production. Water Resour Res 58:E2021wr030297

Yifru BA, Chung I-M, Kim M-G, Chang SW (2022) Augmenting freshwater availability in mountain headwater streams: assessing the sustainability under baseline and future climate change scenarios. Int Soil Water Conserv Res 10:293–307

Yuan L, Lu Y, Liu J, Liu H, Lu Y, Zhou X (2023) Advances in ecological and environmental effects of mountain river sediment. proceedings of pianc smart rivers 2022: green waterways and sustainable navigations. Springer

Zan Y, Gao Y, Jiang Y, Pan Y, Li X, Su P (2022) The effects of lake level and area changes Of Poyang lake on the local weather. Atmosphere 13:1490

Zhao Y, Zhao M, Qi L, Zhao C, Zhang W, Zhang Y, Wen W, Yuan J (2022) Coupled relationship between soil physicochemical properties and plant diversity in the process of vegetation restoration. Forests 13:648

Chapter 21
Evaluation of Innovation and Entrepreneurship Education Based on Fuzzy Comprehensive Evaluation: A Case Study of Water Management

Jie Dai, Haoran Wei, Hengbin Yin, Chengqing Su, Jian Chu, and Yan Tu

Abstract Entrepreneurship related to water environment governance is an important way to promote sustainable development. Water Management majors in some Chinese universities have set up courses on innovation and entrepreneurship to organize students to carry out innovation and entrepreneurship practice. But it is necessary to evaluate the effects of such education. Based on fuzzy comprehensive evaluation (FCE), this paper evaluates the performance of innovation and entrepreneurship education in water environmental management major in universities. Firstly, the fuzzy evaluation index was selected, and the single factor matrix was set by calculating the evaluation weight. In case of a college, the FCE evaluation model was constructed, and the bidirectional top-level processing was used to realize the performance evaluation. The test results show that: the organizational leadership and curriculum content of the case college are good, and other indicators such as success conversion rate are in the average level. It is believed that more efforts should be made for colleges to improve the entrepreneurial environment.

Keywords Entrepreneurship · Water governance · Fuzzy · AHP · FCE

J. Dai · H. Wei (✉)
Institute for Advanced Studies, University of Malaya, Kuala Lumpur, Malaysia
e-mail: s2031355@siswa.um.edu.my

H. Wei
Taizhou Branch, China National Offshore Oil Corporation (CNOOC), Beijing, China

H. Yin (✉) · C. Su
School of Finance and Trade, Wenzhou Business College, Wenzhou, China
e-mail: 20190251@wzbc.edu.cn

J. Chu
School of Arts and Media, Suqian University, Suqian, China

Y. Tu
College of Art, Zhengzhou University of Science and Technology, Zhengzhou City, China

H. Xu (ed.), *Proceedings of the 5th International Symposium on Water Resource and Environmental Management*, Environmental Science and Engineering,
https://doi.org/10.1007/978-3-031-31289-2_21

21.1 Introduction

The *Implementation Plan of National Vocational Education Reform* is a policy put forward by China to guide higher vocational schools to cultivate high-quality technical and skilled personnel (Liu and Hardy 2021). This is the hope that vocational college students in the future can focus on serving enterprises, especially the R&D and product upgrading in small, medium and micro enterprises. In the plan, the importance of expertise in technology R&D and product upgrading is mentioned, as well as the cultivation of innovative and entrepreneurial thinking and ability. Innovation and entrepreneurship education (IEE) aims to cultivate students' innovative thinking ability to cope with the new demands under the background of scientific and technological progress (Jónsdóttir and Macdonald 2019). Innovation and entrepreneurship education is currently regarded as an important way for higher vocational colleges to improve the training quality of highly skilled personnel (Ratten and Jones 2021).

Some regions in China, such as Zhejiang Province, hope to drive professional reform through IEE education reform. For example, it plans to improve the quality of talent training through IEE education reform, so as to provide technical skills support for the constantly upgrading water industry (Bühler et al. 2022). At present, many higher vocational colleges have set up the course "Innovation and Entrepreneurship" for the major of Water Management, and actively guide students to participate in innovation and entrepreneurship practice, such as Internet + College students Innovation and Entrepreneurship Competition, National Higher Vocational Water Technology Skills Competition (Lei et al. 2020). However, compared with the relevant requirements put forward in the theory of innovation and entrepreneurship, the content and methods of IEE education for water governance are relatively simple at present. Some scholars believe that the teaching and practice of innovation and entrepreneurship courses of water environment monitoring and governance major are not well connected with the actual production of enterprises in the industry, without introducing new knowledge, new processes and new technologies in the water industry, resulting in a disconnect between the content of IEE education and enterprises (Dyllick and Muff 2016; Chen et al. 2022, 2021; Chen and Wu 2021). On the one hand, it is difficult to arouse students' passion for learning; On the other hand, it also makes students' innovation and entrepreneurship awareness and ability training free from the rapid development of smart water era.

In recent years, IEE education in the field of environmental protection has received much attention. Its purpose is to cultivate students' environmental awareness and innovation ability through education, so as to promote the development of environmental protection industry. The literature on IEE education quality evaluation in the field of environmental protection caught our attention. Some studies have used questionnaires to assess the quality of IEE education in the field of environmental protection. For example, Li (2018) conducted a questionnaire survey among universities in China to assess the status and quality of IEE education in environmental protection. For example, Bao and Li (2021) used Analytic Hierarchy Process (AHP), principal component analysis (PCA) and data envelopment analysis (DEA) to evaluate the quality of IEE education in environmental protection in Chinese universities. They take educational quality evaluation indicators (such as teacher level, curriculum setting, practical teaching, etc.) as different levels of hierarchy, and obtain the weights of different indicators through questionnaire survey. Then, they used PCA to comprehensively evaluate different indicators to evaluate the quality of IEE education in the field of environmental protection in various universities. The results show that teacher level and practical teaching are the key factors affecting the quality of innovative education in environmental protection. In addition, some studies use data-mining based methods to assess the quality of IEE education. Similarly, Wu et al. (2021) used cluster analysis and decision tree analysis to assess the ability and performance of Taiwanese students in IEE education. By analyzing students' test data and questionnaire data, they divided students into different groups and studied the performance of different groups.

Past research usually studied on environment-related social or educational problems from econometric and socio-psychological perspectives (Lin et al. 2020; Zhang et al. 2022; Chou et al. 2020; Wei et al. 2022; Zhu et al. 2021; Huang et al. 2022; Qian et al. 2021), but IEE education evaluation based on fuzzy theory has been more and more recognized in recent years. This is because the evaluation pertinence of fuzzy comprehensive evaluation method is relatively stronger. Under complex environment, it can minimize the evaluation error and provide reference for the rating method. Therefore, this paper takes a vocational college in Zhejiang Province, China as an example to make an exploratory evaluation.

21.2 Methodology

As shown in Table 21.1, the evaluation framework is divided into three levels: A, B and C. The evaluation starts from the two levels of students and schools and is then divided into four aspects. B = {B1, B2, B3, B4} = {entrepreneurship ability, innovation effect, faculty resources, reputation}. There are nine evaluation indexes C = {C1, C2, C3, C4, C5, C6, C7, C8, C9}. The evaluation system of our study is divided into four levels, the specific evaluation set is V = {excellent, good, medium, poor}.

The correctness of membership degree and membership function has direct influence on the reliability of evaluation results. In this paper, the membership function of each index is determined by logical inference assignment method according to the relevant data of case college. As shown in Table 21.2, according to the specific situation of A university, the membership degree of each indicator is obtained after expert scoring.

Table 21.1 Evaluation framework

First-Level Indicators (A)	Weight		Second-Level Indicators (B)	Weight		Third-Level Indicators (B)	Weight	
Student Level (A1)	0.7	a1	Entrepreneurship ability (B1)	0.6	b1	Organizational leadership (C1)	0.4	c_1
						Operation and management ability (C2)	0.4	
						Social skill (C3)	0.2	
			Innovation effect (B2)	0.4		Entrepreneurial achievement (C4)	0.6	c_2
						Success rate (C5)	0.4	
School Level (A2)	0.3		Faculty resources (B3)	0.5	b_2	Teacher background (C6)	0.5	c_3
						Course content (C7)	0.5	
			Reputation (B4)	0.5		Entrepreneurial environment (C8)	0.6	c_4
						School reputation (C9)	0.4	

21.3 Results

The weight distribution set of first-level indicators is:

$$a1 = [(0.70.3)]$$

The weight distribution set of the secondary indicators is:

$$b1 = [(0.60.4)]$$
$$b2 = [(0.50.5)]$$

The weight distribution set of the third-level indicators is:

$$c1 = [(0.40.40.2)]$$
$$c2 = [(0.60.4)]$$
$$c3 = [(0.50.5)]$$
$$c4 = [(0.60.4)]$$

Table 21.2 Statistical table of expert evaluation

Expert evaluation items (d)		Excellent	Good	Medium	Poor
Organizational leadership (C1) Operation and management ability (C2)	Score	7	3	0	0
	Normalization	0.7	0.3	0	0
Social skill (C3) Entrepreneurial achievement (C4)	Score	5	4	1	0
	Normalization	0.5	0.4	0.1	0
Success rate (C5) Teacher background (C6)	Score	2	5	2	1
	Normalization	0.2	0.5	0.2	0.1
Course content (C7) Entrepreneurial environment (C8)	Score	3	4	3	0
	Normalization	0.3	0.4	0.3	0
School reputation (C9) Organizational leadership (C1)	Score	3	5	2	0
	Normalization	0.3	0.5	0.2	0
Operation and management ability (C2) Social skill (C3)	Score	5	4	0	1
	Normalization	0.5	0.4	0	0.1
Entrepreneurial achievement (C4) Success rate (C5)	Score	6	3	1	0
	Normalization	0.6	0.3	0.1	0
Teacher background (C6) Course content (C7)	Score	5	3	2	0
	Normalization	0.5	0.3	0.2	0
Entrepreneurial environment (C8)	Score	5	3	1	1
	Normalization	0.5	0.3	0.1	0.1

Based on the judgment of 10 experts on each evaluation index, the three-level expert evaluation statistical table of each level is obtained (Table 21.2).

The evaluation set of entrepreneurship ability (B1):

$$\text{d1} = [\text{C1 C2 C3}]^{\text{T}} = \begin{bmatrix} 0.7 & 0.3 & 0 & 0 \\ 0.5 & 0.4 & 0.1 & 0 \\ 0.2 & 0.5 & 0.2 & 0.1 \end{bmatrix}$$

The evaluation set of innovation effect (B2):

$$\text{d2} = [\text{C4 C5}]^{\text{T}} = \begin{bmatrix} 0.3 & 0.4 & 0.3 & 0 \\ 0.3 & 0.5 & 0.2 & 0 \end{bmatrix}$$

The evaluation set of faculty resources (B3):

$$\text{d3} = [\text{C6 C7}]^{\text{T}} = \begin{bmatrix} 0.6 & 0.3 & 0.1 & 0 \\ 0.5 & 0.4 & 0.0 & 0.1 \end{bmatrix}$$

The evaluation set of reputation (B4):

$$d4 = [C8\ C9]^T = \begin{bmatrix} 0.5 & 0.3 & 0.2 & 0 \\ 0.5 & 0.3 & 0.1 & 0.1 \end{bmatrix}$$

The third-level fuzzy comprehensive evaluation results are calculated based on the formula: weight * evaluation set, therefore, then:

$$B1 = c1*d1 = \begin{bmatrix} 0.4 & 0.4 & 0.2 \end{bmatrix} * \begin{bmatrix} 0.7 & 0.3 & 0 & 0 \\ 0.5 & 0.4 & 0.1 & 0 \\ 0.2 & 0.5 & 0.2 & 0.1 \end{bmatrix} = \begin{bmatrix} 0.52 & 0.29 & 0.08 & 0.02 \end{bmatrix}$$

$$B2 = c2*d2 = \begin{bmatrix} 0.6 & 0.4 \end{bmatrix} * \begin{bmatrix} 0.3 & 0.4 & 0.3 & 0 \\ 0.3 & 0.5 & 0.2 & 0 \end{bmatrix} = \begin{bmatrix} 0.30 & 0.44 & 0.24 & 0.02 \end{bmatrix}$$

$$B3 = c3*d3 = \begin{bmatrix} 0.5 & 0.5 \end{bmatrix} * \begin{bmatrix} 0.6 & 0.3 & 0.1 & 0 \\ 0.5 & 0.4 & 0.0 & 0.1 \end{bmatrix} = \begin{bmatrix} 0.55 & 0.35 & 0.05 & 0.05 \end{bmatrix}$$

$$B4 = c4*d4 = \begin{bmatrix} 0.6 & 0.4 \end{bmatrix} * \begin{bmatrix} 0.5 & 0.3 & 0.2 & 0 \\ 0.5 & 0.3 & 0.1 & 0.1 \end{bmatrix} = \begin{bmatrix} 0.50 & 0.30 & 0.16 & 0.04 \end{bmatrix}$$

The second-level fuzzy comprehensive evaluation results are therefore:

$$A1 = b1*B12 = \begin{bmatrix} 0.6 & 0.4 \end{bmatrix} * \begin{bmatrix} 0.52 & 0.29 & 0.08 & 0.2 \\ 0.30 & 0.44 & 0.24 & 0.02 \end{bmatrix} = \begin{bmatrix} 0.43 & 0.35 & 0.14 & 0.13 \end{bmatrix}$$

$$A2 = b2*B34 = \begin{bmatrix} 0.5 & 0.5 \end{bmatrix} * \begin{bmatrix} 0.55 & 0.35 & 0.05 & 0.05 \\ 0.50 & 0.30 & 0.16 & 0.04 \end{bmatrix} = \begin{bmatrix} 0.78 & 0.33 & 0.11 & 0.05 \end{bmatrix}$$

The first-level fuzzy comprehensive evaluation results are therefore:

$$A = a1*A12 = \begin{bmatrix} 0.7 & 0.3 \end{bmatrix} * \begin{bmatrix} 0.43 & 0.35 & 0.14 & 0.13 \\ 0.78 & 0.33 & 0.11 & 0.05 \end{bmatrix} = \begin{bmatrix} 0.54 & 0.35 & 0.13 & 0.13 \end{bmatrix}$$

According to the principle of maximum membership, it shows that the IEE education in the case vocational colleges is evaluated as "excellent".

21.4 Conclusion

Innovation and entrepreneurship education reform is considered to be an inevitable requirement for water management major to cultivate talents to adapt to the transformation and upgrading of water industry. Through the comprehensive evaluation of IEE education quality in the case college, it is found that the organizational leadership and curriculum content of the case colleges are good, and other indicators such as success conversion rate are in the average level. But overall, the school's rating is excellent, which indicates that the case school can be used as a good reference standard to judge the performance of other universities. However, in order to improve the quality of IEE education for college students, we believe that the case of higher vocational colleges should appropriately improve the entrepreneurial environment, and at the same time provide more soft support for entrepreneurship.

Acknowledgments Thanks to the support from "Fengcheng Talent Program" of Taizhou Association for Science and Technology.

References

Bao X, Li F (2021) A methodology for supplier selection under the curse of dimensionality problem based on fuzzy quality function deployment and interval data envelopment analysis. PLoS ONE 16(7):e0253917

Bühler MM, Jelinek T, Nübel K (2022) Training and preparing tomorrow's workforce for the fourth industrial revolution. Educ Sci 12(11):782

Chen M, Wu X (2021) Attributing academic success to giftedness and its impact on academic achievement: the mediating role of self-regulated learning and negative learning emotions. Sch Psychol Int 42(2):170–186

Chen M, Xu Q, Cui M (2021) Sleep problems, attention, and classroom learning behaviors of Chinese elementary school children: the moderating role of gender. Sch Psychol Int 42(4):341–357

Chen M, Zee M, Roorda DL (2022) Assessing student–teacher relationship quality in cross-cultural contexts: psychometric properties of student–teacher relationship drawings. Eur J Dev Psychol 19(5):770–784

Chou LC, Zhang WH, Wang MY, Yang FM (2020) The influence of democracy on emissions and energy efficiency in America: new evidence from quantile regression analysis. Energy Environ 31(8):1318–1334

Dyllick T, Muff K (2016) Clarifying the meaning of sustainable business: introducing a typology from business-as-usual to true business sustainability. Organiz Environ 29(2):156–174

Huang W, Dai J, Xiong L (2022) Towards a sustainable energy future: factors affecting solar-hydrogen energy production in China. Sustain Energy Technol Assess 52:102059

Jónsdóttir SR, Macdonald MA (2019) The feasibility of innovation and entrepreneurial education in middle schools. J Small Bus Enterp Dev 26(2):255–272

Lei H, Zuo J, Deng H (2020) Research and practice on the "Three Cooperation" mode of personalized and Elite Talents training in higher vocational colleges. Educ Reform Dev 2(2)

Li Y (2018) Study of the effect of environmental education on environmental awareness and environmental attitude based on environmental protection law of the People's Republic of China. Eurasia J Math, Sci Technol Educ 14(6):2277–2285

Lin HC, Chou LC, Zhang WH (2020) Cross-Strait climate change and agricultural product loss. Environ Sci Pollut Res 27:12908–12921

Liu S, Hardy I (2021) Understanding Chinese national vocational education reform: a critical policy analysis. J Vocat Educ Train 1–23

Qian X, Bai Y, Huang W, Dai J, Li X, Wang Y (2021) Fuzzy technique application in selecting photovoltaic energy and solar thermal energy production in belt and road countries. J Energy Stor 41:102865

Ratten V, Jones P (2021) Covid-19 and entrepreneurship education: implications for advancing research and practice. Int J Manag Educ 19(1):100432

Wei H, Dai J, Maharik I, Ghasemi A, Mouldi A, Brahmia A (2022) Simultaneous synthesis of H_2, O_2, and N_2 via an innovatory energy system in coronavirus pandemic time: design, techno-economic assessment, and optimization approaches. Int J Hydrog Energy 47(62):26038–26052
Wu SF, Lu YL, Lien CJ (2021) Detecting students' flow states and their construct through electroencephalogram: reflective flow experiences, balance of challenge and skill, and sense of control. J Educ Comput Res 58(8):1515–1540
Zhang WH, Chou LC, Chen M (2022) Consumer perception and use intention for household distributed photovoltaic systems. Sustain Energy Technol Assess 51:101895
Zhu X, Dai J, Wei H, Yang D, Huang W, Yu Z (2021) Application of the fuzzy optimal model in the selection of the startup hub. Discr Dyn Nat Soc 1–9